Dryland Agriculture and Wasteland Management

Emerging Issues and Extension Strategies

Edited by

C. Karthikeyan
K. Thangaraja
C. Cinthia Fernandez
K. Chandrakandan

ATLANTIC

PUBLISHERS & DISTRIBUTORS (P) LTD

Published by

ATLANTIC

PUBLISHERS & DISTRIBUTORS (P) LTD

7/22, Ansari Road, Darya Ganj,
New Delhi-110002
Phones : +91-11-23273880, 23275880, 23280451
Fax : +91-11-23285873
Web : www.atlanticbooks.com
E-mail : info@atlanticbooks.com

Branch Office
5, Nallathambi Street, Wallajah Road,
Chennai-600002
Phones : +91-44-64611085, 32413319
E-mail : chennai@atlanticbooks.com

ISBN 978-81-269-1099-1

Printed in India at Nice Printing Press, A-33/3A, Site-IV,
Industrial Area, Sahibabad, Ghaziabad, U.P.

Preface

Dryland agriculture in India is always a challenge, since crop production in dryland depends on monsoon showers. Out of the net sown area of 136.18 m ha, the dryland accounts for 93.13 m ha (68.4%) and contributes 55 million tonnes of foodgrain production. Dryland agriculture on 93 m ha supports 40 per cent of India's human population, 60 per cent of cattle heads and contributes 44 per cent to total food production in the country. Dry farming regions contribute considerably to total production of millet, pulses, oilseeds and fibre, since large shares of area under these crops are rainfed. About 42 per cent of total foodgrain production of the country is received from dry farming regions. In order to meet the targeted foodgrain production of 240 million tonnes, production from drylands has to be increased from 60 million tonnes to about 144 million tonnes.

Crop selection for the dryland region, along with its management, has a major impact on how efficiently the available natural resources are utilized to produce an economic yield. Though several factors are affecting crop production in drylands, rainfall is the most important factor recognized, which is highly unpredictable in terms of onset and duration of monsoon and distribution of rainfall across the country. Unfavourable crop growth environment, limited choice of crops and varieties, poor soils, low cropping intensity, short growing season and low unstable productivity are the major concerns of dryland farmers.

There is an urgent need to formulate strategies to increase agricultural production in the dry farming tracts. Dry farming is purely based on distribution of rainfall. Usually, in dryland areas, farmers are able to get single crop. Crop failure is a recurring problem in arid and semi-arid tracts. Farmers are generally following mixed cropping system to avoid crop failure. Normally, cereals and pulses are grown as mixed crops. One or two animals are also domesticated by them to get additional income. Proper utilization of the land area will give stable income and lead to sustainable development of dryland farmers. The low productivity of agriculture in dry farming regions is due to the cumulative effect of many constraints for crop production. Such constraints need to be identified to overcome with suitable remedies.

The realisation of the millennium development goals, which focus on poverty alleviation, has given a new urgency to agricultural policy. For many years government debated the merit of investing in low productive areas, i.e. wastelands. It has been estimated that about two-thirds of the cultivable area in our country remain as drylands and about 20 per cent is left as wastelands, due to various reasons.

Wasteland management planning in developing countries is usually hampered by limited manpower resources and time. Hence, wasteland technologies must be compatible with socio-economic concerns in rural development. Wasteland technologies will gain acceptance among the rural masses only when these basic issues are adequately addressed.

Development of wasteland is an important programme which includes checking land degradation, increasing bio-mass availability and putting wastelands to use. Sustainable use of the wasteland is possible through implementation of various development projects and allied farming activities.

This book is an outcome of the national seminar on "Emerging Issues and Development Strategies for Dryland Agriculture and Wasteland Management", held at CRIDA, Hyderabad during 6-8 March 2006. The seminar, *inter alia,* suggested ways to find out technological strategies and disseminate them to ensure wide application thereof.

The topics covered are related to dryland agriculture and wasteland management with respect to soil management in dryland areas; different farming systems and extension strategies in dry tracts; indigenous technical knowledge applied in dryland area; watershed management and sustainability in different areas.

The articles in this anthology have been contributed by scientists, scholars and researchers in the field of dryland agriculture. We have edited the papers with utmost care to make them lucid and more presentable. The book will prove to be an ideal reference source for students, researchers, administrators and those concerned with the development of agriculture in India.

We are thankful to all the contributors for their intellectual endeavours. We are grateful to Dr. K.R. Gupta, Honourable Advisor, Atlantic Publishers and Distributors (P) Ltd. for publishing the book.

C. Karthikeyan
K. Thangaraja
C. Cinthia Fernandez
K. Chandrakandan

Contents

Identification of Problems in an Agro-Ecosystem Through Snowball Technique

1

L.M. Ahire, Sandhya Shenoy and
V.K. Jayaraghavendra Rao

Introduction

Need-based research and development of locally suitable technologies and their adoption through the participatory approach and group action facilitated by extension functionaries and policy-makers is essential to integrate farmer's needs and aspirations in appropriate technology development and transfer.

The paper deals with agro-eco situation in Reddypalem village of Mahboobnagar district and analysis of problems in village through Snowball technique using frequency and rank order, thereby analysing the situation and identifying commonly acceptable remedies for formulating time-bound action plans.

Methodology

The study was conducted during the month of November 2003, to know the agro-ecosystem of the village and to analyse problems faced by the farmers by using Participatory Rural Appraisal tools/steps. After introduction with the key informants/ villagers, a transect walk was conducted to get first hand information about village in terms of land utilisation, cropping pattern, livestock, natural and physical resources, prominent features of village, problems and opportunities in village. This lead to in-depth understanding regarding the issues involved.

The ecological understanding of village will help in planning regional, rural and agricultural development programmes with less negative environmental effects (Ryszkowski, 1989). The farmers of Reddypalem village discussed on different aspects related to crops, animals and resources. The other information related to the system was also collected while constructing the agro-ecological map of the village.

Seasonal analysis explores seasonal problems and opportunities by diagramming changes, month by month throughout the year (Paul and Pierre, 1995). By asking the season-wise activities of farmers seasonal calendar was drawn.

Preference ranking has been recommended to know the farming problems by Theis and Grady (1991). In preference ranking, a farmer or a key informant is asked to list the problems of farm in the village and indicate which problem he prefers to place as the top first, which problem as second, third and fourth, etc.

The ranks are scored in the reverse order, i.e. rank 1 having highest score, rank 2 the next score and so on. The scores of all ranks are added and the total score was considered for arriving at overall rank of the problem, the highest score being the topmost priority problem.

Problem tree for getting to the root of the problem facilitates the identification of the possible interventions. It helps in decision-making process for the most acceptable strategy in order to solve the problem (Fussel, 1995).

Results and Discussion

The village transect reveals that Reddypalem falls under Southern Telangana Zone of Andhra Pradesh. This zone consists of five districts of Andhra Pradesh (12° 14′-19° 54′ N latitude, 76° 50′-84° 50′ E longitude). The village lies under semi-arid tropical area where the mean annual rainfall ranges from 750 mm to 1000 mm, the temperature ranges from 16.5°C to 48.8°C. The basic information and agro-ecosystem information reveal that the farming families are homogeneous with a single community (Christian) and are 75 in number indicating the small size of the village and primary occupation is agriculture.

All the 75 farm households are medium farmers. The village transect gives a clear picture of the land use pattern (see transect map). Farmers having different types of lands and its uses under each type of land and problems faced by them were discussed with key informants and noted on the map.

Village: Reddypalem **Village Panchyat:** Mallapur

Mandal: Kothur

District: Mahboobnagar **State:** Andhra Pradesh

Number of families	:	75
Community	:	Homogeneous (Christian only)
Total population	:	347
Male and female	:	250
Number of children (Below 18 years)	:	97
Occupational distribution	:	Only agriculture

Agricultural Scenario in the Village

Total cultivated land	:	950 acres
Average land holding per family	:	12.6 acres
Average land holding per head	:	2.7 acres

Crop Husbandry

Maize	300 acres
Cotton	150 acres
Paddy	150 acres
Chilli	50 acres
Other crops (redgram, Bengal gram, dolichos beans, tomato, brinjal, etc.)	300 acres

Animal Husbandry

Cows	
Cross breed jersey	43
Desi	26
Buffalo	
Murrah	48
Desi	28
Bullocks (Ongole and desi)	261
Total livestock population	**406**

Village: Reddypalem **Mandal**: Kothur **Dist**. Mahboobnagar

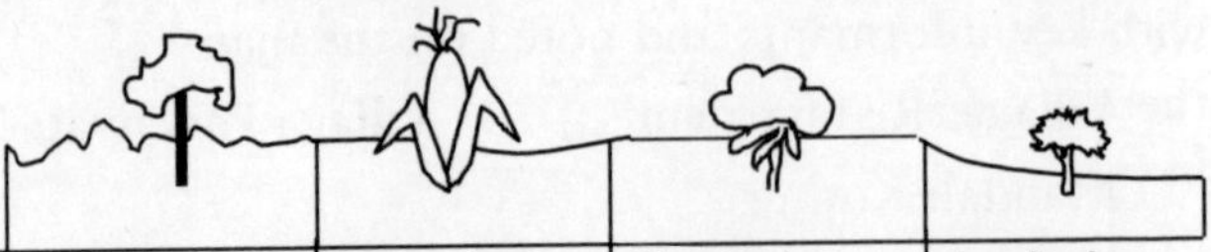

Topography	Undulating	Medium land	Plain land	Low land
Soil characteristics	Gravel with rocks brown in colour	Chalka (red soil brownish with loam)	Chalka (red soil with sandy loam) and Nalla regadi (black soil with fine particles)	Soudu Bhoomi (black saline soils)
Vegetation	Soapnut, Sisso, Casia simea, Subabool, Teak	Coconut, Teak, Neem	Thumma, Peltoforum	Prosoupis julifera, Thumma
Crops	—	Maize, Jowar, Dolichos bean, Bhendi	Maize, Cotton, Chilli, Paddy, Tomato, Brinjal, Bengal gram, Redgram	Paddy, Cotton, Fodder crops
Livestock	—	Bullocks, Buffalo, Cows	Bullocks, Buffalo, Cows, Backyard poultry	—
Weeds	Natural vegetation	Parthenium, Cyprus	Parthenium, Garka, Tridox	Parthenium, Garka, Sisakamma
Water source	—	Rain-fed, Tube well	Tube well, rain-fed	Rain-fed
Problems	—	Untimely/erratic rainfall, Pest/Diseases, Weeds, unavailability of labour	Shortage of fodder, Water scarcity, poor transport, lack of markets, irregular power supply	Water stagnation, Weeds
Opportunities	Development of Silvi-systems	Adoption of IPM practices, water conservation, compost making, use of weedicides	Crop rotation, Orchards plantation, Sheep farming, Development of common grazing land, adoption of bio-pesticides	Development of watershed ponds, soil reclamation

Agro-ecosystem

Agro-eco map reveals the different types of soil locally called Chalka (red soil) were utilised for cultivation of maize, jowar, dolichos beans and bhendi and red soil with sandy loam utilised for cultivation of maize, chilli, redgram and tomato. The soil called Nalla regadi (black soil) and Soudu bhoomi (saline soil) were utilised for cultivation of cotton, paddy and fodder crops, there is also a vegetation of *Prosopis julifera* on *Soudu bhoomi* in eastern part of the village which is mainly used for fuel purpose by the villagers. There is also evidence of development of social forestry in village and on undulating or uplands. The system was observed in the village is crop based agro-ecosystem and totally rain-fed farming pattern in use because of less water resources available, low water table and no governmental intervention for water conservation was observed.

Livestock situation

Compared to buffaloes, cows and bullocks were more in the village used for cultivation, while other animals were used for milch for household. Opportunity exists for cattle and sheep farming in the village through development of common grazing land by reducing vegetation of the *Prosophis julifera.*

The major problem for livestock is lack of water sources, shortage of fodder and diseases like Foot and Mouth (FMD), Black Water (BQ) and *Hemorrhagic septicemia.*

Seasonal calendar

Seasonal calendar reveals that farmers were busy in farming throughout the year except in the period of April-May, due to less work in the farm. Seasonal activities like sowing, weeding, fertilizer application, spraying, harvesting, etc. were carried out during late June to December. Due to erratic rainfall farmers suffer with more incidences of weeds and pests/diseases. There is shortage of labour during the cropping season especially in the months of July-December considering activities like weeding, spraying, harvesting, etc. The other activities taken up by farmers during November to May were feeding of animals, growing of fodder crops, milking, etc. March to May was a harsh period for them because they had to face the water scarcity, shortage of fodder and irregular power supply. The special feature of the village was the homogeneous community (Christian) that mainly celebrates Christmas. Other celebrations were very few for them except marriages during December to May.

Identification of problems

Problems were identified with the help of randomly selected key informants/farmers (30) by asking individually to rank the major problems faced in an agro-ecosystem as they perceived, rank one being the most severe problem. The preference was based on farmer's perception about the severity of the problem compared to other problems. The farmers identified different problems in the village like higher incidence of weeds, high labour wages, pests and diseases in cotton, lack of storage facilities, low market prices, shortage of fodder, lack of financial assistance, high cost of cultivation and unavailability of labour in time.

Based on the rank frequencies and total score(s) obtained, "Pests and diseases of cotton" was the most severe problem in the village as perceived by the farmers, which, was ranked, first by twenty-one farmers (Table 1). This is followed by "more incidence of weeds" which is ranked second. The top-most priority of villagers is to reduce the pests and diseases in cotton as perceived by them. The problem tree was constructed with possible and easily adaptable solutions, by discussing with the farmers and the stakeholders.

Conclusion

Snowball technique using rank frequencies and order is useful to identify technological interventions through agro-eco situation analysis. Four types of topographic lands have been observed in the village. Undulating soils are gravel, brown in colour with rocks. Medium lands are red soils (*Chalka*) brownish with loam, in plain land red soils (*Chalka*) and black soils (*Nalla regadi*) was observed and low lands are black saline soils (*Soudu bhoomi*). These soils influence the cropping system of the village. Based on the frequencies and ranks of the farmers given to the problems in the system pest and diseases of cotton and more incidences of weeds were the major problems in the village. The problem tree and solution tree were developed to reduce the pests and diseases in cotton and necessary suggestions were given to the cotton growers. Therefore, participatory approaches for

Table 1: Rank Frequencies for the Identified Agricultural Problems

S. No.	Problems / Ranks	Frequencies/Ranks of farmers									Rank / overall total score
		1	2	3	4	5	6	7	8	9	
1	More incidence of weeds	3 (27)*	18 (144)	3 (21)	3 (18)	–	–	2 (6)	–	1 (1)	II (217)
2	High labour wages	–	1 (8)	2 (14)	2 (12)	5 (25)	18 (72)	1 (3)	1 (2)	–	VI (136)
3	Pest and diseases of cotton	21 (189)	3 (24)	–	2 (12)	–	–	–	1 (2)	3 (3)	I (230)
4	Lack of storage facilities	–	1 (8)	–	1 (6)	3 (15)	3 (12)	17 (51)	3 (6)	2 (2)	VII (100)
5	Low market prices	2 (18)	1 (8)	2 (14)	–	16 (80)	3 (12)	1 (3)	3 (6)	2 (2)	V (143)
6	Shortage of fodder	–	3 (24)	–	2 (12)	1 (5)	1 (4)	2 (6)	3 (6)	18 (18)	IX (75)
7	Lack of financial assistance	–	1 (8)	3 (21)	1 (6)	2 (10)	4 (16)	2 (6)	14 (28)	3 (1)	VIII (96)
8	High cost of cultivation	3 (27)	1 (8)	2 (14)	18 (108)	2 (10)	2 (8)	2 (6)	–	–	III (181)
9	Unavailability of labour in time	1 (9)	2 (16)	17 (119)	1 (6)	2 (10)	–	3 (9)	2 (4)	2 (2)	IV (175)

*Figures in parentheses indicate score(s).

technological interventions is necessary to integrate farmers needs and aspirations.

REFERENCES

Conway, G.R. (1985). Agro-ecosystem Analysis Agricultural Administration 20: 31-35.

Conway, G.R. (1987). The Properties of Agro-ecosystems. *Agricultural Systems*. 24: 95-117.

Mettrick, H. and Wessel, M. (1986). Farming System Research as an Analytical Framework and Tool for Training. Bulletin, 23, ICRA, Wageningen, The Netherlands.

Rao, V.K.J. and Sandhya Shenoy, N. (2005). Training Manual on Participatory Rural Appraisal and Action Learning for Research and

Extension in Agriculture, September 22-29, 2005, NAARM, Hyderabad, India, p. 223.

Sabarathnam, V.E. (2002). R/R/PRA (PLA) for Agriculture, Vamsaravath Publishers, Hyderabad, p. 448.

Singh, R.P. (2001). Development Plan of Village Sandhir on the Basis of PRA Techniques. *MANAGE, Extension Research Review*, July-December 2001, pp. 94-107.

Soam, S.K., Raut, A., Anu Sharma, Mishra, D., Manikantan, M.R. and Suman, R.S. (2003). Participatory Agro-Ecosystem Analysis and Identification of Problems of a Village in Bundelkhand Region. *MANAGE, Extension Research Review*, Jan-June 2003, pp. 21-31.

Suryamani, M. (2003). Diagnosis of the General Agricultural Problems of Farmers through Village Level Participatory Approach in Mahboobnagar District of Andhra Pradesh. *MANAGE, Extension Research Review*, July-December 2003, p. 168.

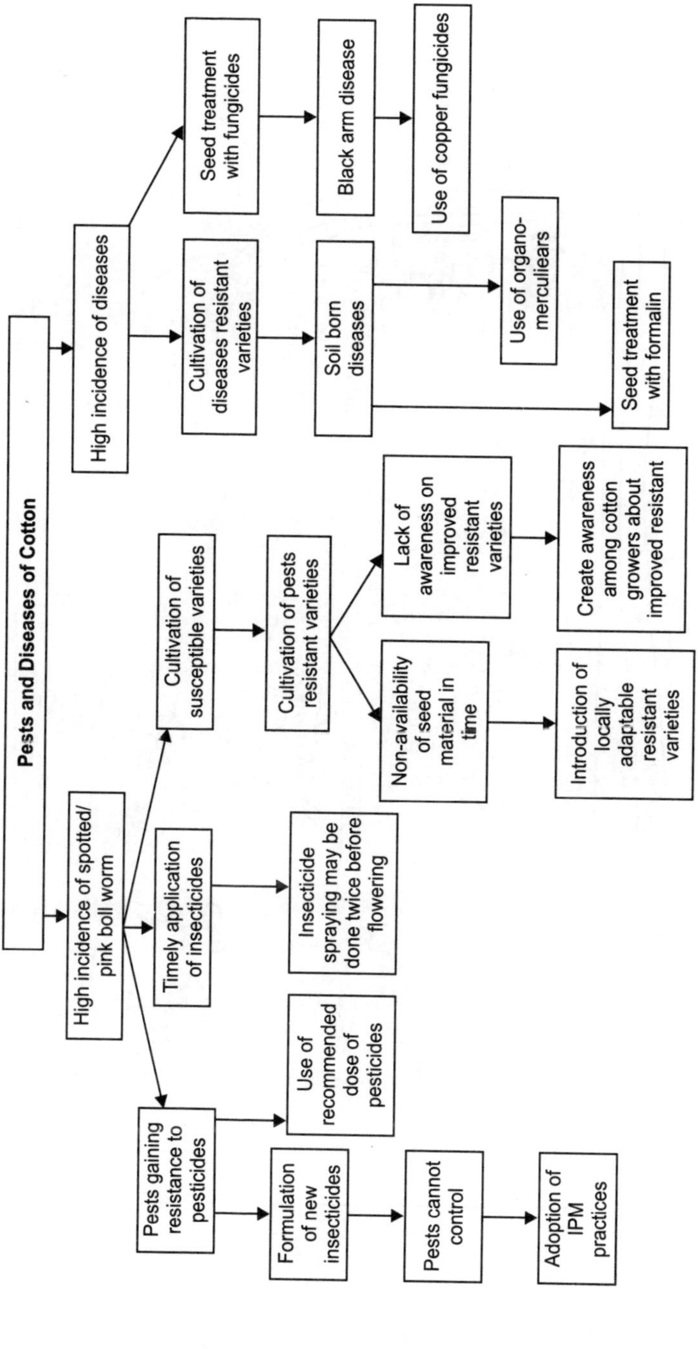

Problem Tree/Solution tree of top most problems in an agro eco-system of cotton.

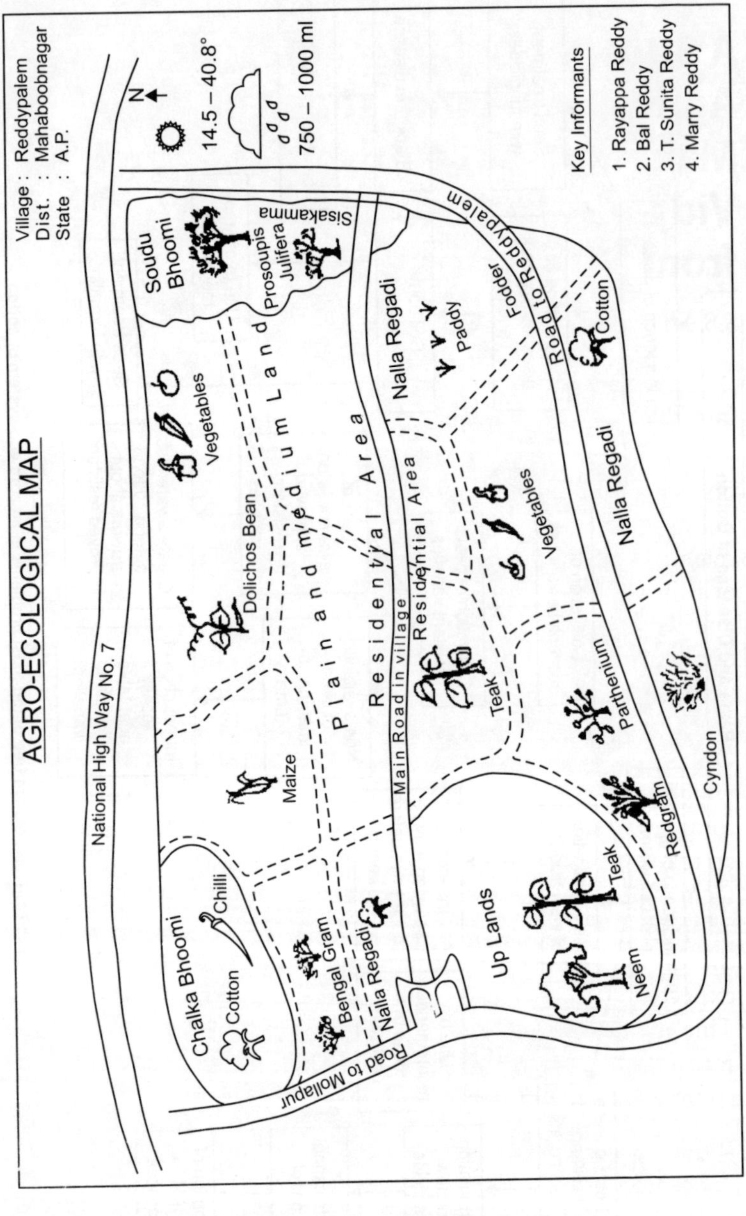
AGRO-ECOLOGICAL MAP
Village : Reddypalem
Dist. : Mahaboobnagar
State : A.P.
N
14.5 – 40.8°
750 – 1000 ml
Key Informants
1. Rayappa Reddy
2. Bal Reddy
3. T. Sunita Reddy
4. Marry Reddy
National High Way No. 7
Soudu Bhoomi
Prosoupis Julifera
Sisakamma
Vegetables
Dolichos Bean
Plain and medium Land
Maize
Chalka Bhoomi
Chilli
Cotton
Bengal Gram
Nalla Regadi
Road to Mollapur
Residential Area
Main Road in village
Residential Area
Teak
Nalla Regadi
Paddy
Fodder
Road to Reddypalem
Cotton
Vegetables
Nalla Regadi
Parthenium
Cyndon
Redgram
Teak
Up Lands
Neem

A Study of Adoption of Dryland Agricultural Innovations of Mahatma Phule Krishi Vidyapeeth by the Farmers from Maharashtra State

2

S.B. Shinde and *A.G. Sawant*

Introduction

The intensive research and extension efforts focused mainly on irrigated land till date but much of the needed attention is not given to rainfed agriculture. The commitment of agricultural scientists and other supporting agencies to face the challenges of this country lies in the extensive management of plant, soil and water resources with an intensive focus on improving the quality of rainfed agricultural production system. The efforts in this regard have been undertaken by the farm scientists working at Dryland Research Station, Solapur, Maharashtra. Only 16 per cent area has irrigation facilities in Maharashtra which cannot be increased to more than 25 per cent of the total cropped area. Solapur district lies in the heart of drought-stricken region of the country though the dry farming technology was evolved at the Dryland Research Station, its adoption was not to fullest extent. The per hectare yields of the dryland crops on farmers' fields are low than the yield, expected by adopting the improved dryland farming practices. This might be due to non-adoption of the improved dryland farming practices. The dryland farming might also be facing some constraints in adopting these practices.

Objectives of the Study

1. To study the personal, socio-economic and psychological characteristics of the dryland farmers.

2. To study the level of knowledge of dryland farmers about the selected dryland agricultural innovations.
3. To study the extent of adoption of selected dryland agricultural innovations by the farmers.
4. To study the constraints faced by the dryland farmers in adoption of selected dryland agricultural innovations and their suggestions for overcoming the constraints.

Methodology

The study was conducted in 10 randomly selected villages viz., Nandur, Arali, Borale, Salgar Bk., Dhonaj, Hivargaon, Kagesti, Lendave Chincholi and Uchethan of Mangalwedha tahsil from Solapur district of Maharashtra State. From these selected villages 130 samples were drawn by using the systematic random sampling techniques. The data were collected personally by interviewing the respondents. The variables considered for the study were age, education, size of family, land holding, annual income, social participation, source of information, risk orientation, cosmopoliteness, knowledge and adoption. The statistical tools like mean, standard deviation and correlation coefficient were applied for analysis of the data.

Results and Discussion

The results of the investigation are presented in the following order: Personal, Socio-economic and Psychological characteristics of the respondents (Table 1).

Table 1: Distribution of the Dryland Farmers According to their Personal, Socio-economic and Psychological Characteristics

Sr. No.	Category	Frequency	Per cent
	Age:		
1	Young (Age up to 31)	24	18.50
2	Middle Age (32 to 55)	89	68.50
3	Old (56 and above)	17	13.00
	Education:		
1	Illiterate (No formal education)	8	6.20
2	Primary education (1st to 7th std.)	64	49.10
3	Secondary education (8th to 12th std.)	39	30.00
4	Higher education (above 12th std.)	19	14.70

Sr. No.	Category	Frequency	Per cent
	Social participation:		
1	Low (up to score 3)	43	33.10
2	Medium (score 4 to 7)	6	4.30
3	High (score 8 and above)	81	62.40
	Size of family:		
1	Small (up to 3 member)	39	30.00
2	Medium (4 to 7 member)	79	60.80
3	Large (8 and above member)	12	9.20
	Size of land holding:		
1	Small (score up to 2 ha)	95	73.08
2	Medium (score 2.01 to 4 ha)	30	23.08
3	Large (score 4.01 and above ha)	5	3.84
	Annual income:		
1	Low (up to Rs. 22000)	22	17.00
2	Medium (Rs. 22001 to 52000)	89	68.40
3	High (Rs. 52001 and above)	19	14.60
	Cosmopoliteness:		
1	Low (score up to 6)	6	4.60
2	Medium (score 7 to 17)	15	11.50
3	High (score 18 and above)	109	83.80
	Risk orientation:		
1	Low (score up to 21)	22	17.00
2	Medium (score 22 to 30)	17	13.00
3	High (scoer 31 and above)	91	70.00
	Level of farm information source use:		
1	Low (score up to 19)	22	17.00
2	Medium (score 20 to 29)	87	67.00
3	High (score 30 and above)	21	16.00

It was observed that majority of the dryland farmers were middle aged, primary educated, belonged to medium sized families. They were having low to medium levels of annual income, medium level use of farm information sources.

Information source use

The findings regarding information source use for getting knowledge about dryland agricultural innovations by the dryland farmers are presented in Table 2.

Table 2: Distribution of the Dryland Farmers according to their Information Source Use

Sr. No.	Sources of farm information	No. of farmers	Per cent
I	**Interpersonal sources**		
A	**Personal localite**		
1	Progressive farmers	79	60.70
2	Friends	93	71.50
3	Neighbours	59	43.30
4	Relatives	68	52.33
5	Family	37	28.40
B	**Personal cosmopolite sources**		
1	Gramsevak	63	48.40
2	Agricultural officers	36	27.60
3	Agricultural extension officers	81	62.30
4	Village extension workers	31	23.80
II	**Mass media sources**		
A	**Print media**		
1	Newspapers	39	30.00
2	Farm magazines	27	20.70
3	Farm diaries	16	12.30
B	**Electronic media**		
1	Radio	52	40.00
2	Television	41	31.50

The data presented in Table 2 indicates that the dryland farmers predominantly used friends, progressive farmers and relatives as personal localite sources of farm information, Agricultural Extension Officers and Gramsevaks as personal cosmopolite sources. Amongst the electronic and print media sources a considerable proportion of them used Radio, TV and Newspapers followed by the farm magazines.

Knowledge level about dryland agricultural innovations

The information possessed by the dryland farmers in respect of dryland agricultural innovation is given in Table 3.

Table 3: Distribution of Farmers according to their Level of Knowledge of Dryland Agricultural Innovations

Sr.No.	Category	Frequency (N=130)	Per cent
1	Low (score up to 17)	25	19.20
2	Medium (score 18 to 25)	79	60.80
3	High (score 26 and above)	26	20.00

It is observed from Table 3 that majority (60.80 per cent) of the dryland farmers had medium level of knowledge. While 20.00 per cent had high level knowledge and 19.20 per cent of them had low level of knowledge of dryland agricultural innovations.

Extent of adoption of selected dryland agricultural innovations by the farmers

The information regarding extent of adoption of selected dryland agricultural innovation by the farmers is shown in Table 4.

Table 4: Distribution of the Farmers according to their Extent of Adoption of Dryland Agricultural Innovations

Sr. No.	Category	Frequency (N=130)	Per cent
1	Low level adoption (adoption index up to 50)	26	20.00
2	Medium level adoption (adoption index from 51 to 80)	81	62.31
3	High level adoption (adoption index 81 and above)	23	17.69

The data shown in Table 4 reveal that a large majority of the dryland farmers had medium level of adoption of selected dryland agricultural innovations (62.31 per cent). Only 17.69 per cent of them had higher level of adoption and 20.00 per cent had low level of adoption of dryland agricultural innovations.

Practicewise knowledge and adoption of selected dryland agricultural innovations by the farmers

The information regarding knowledge and adoption of selected dryland agricultural innovations by the farmers is shown in Table 5.

The data shown in Table 5 revealed that majority of the dryland farmers had partial to complete knowledge about the dryland agricultural innovations except live bunding and use of mulching. It was observed that less than 40 per cent of the farmers were having complete knowledge about the dryland agricultural innovations viz., levelling of land, contour bunding, contour cultivation, live bunding, use of mulching, crop planning on the basis of soil depth, use of drought resistant varieties, strip cropping, inter-cropping, use of protective irrigation and optimum use of chemical fertilizers. Whereas regarding extent adoption it is observed that only dryland agricultural innovations viz., contour cultivation ploughing once in three years, timely sowing and weed control have more than 40 per cent of the farmers have complete knowledge about dryland agricultural innovations.

Table 5: Distribution of Farmers according to Their Practice-wise Knowledge and Adoption of Selected Dryland Agricultural Innovations

Sr. No.	Dryland agricultural innovations	Complete		Partial		No	
		Know-ledge	Adoption	Know-ledge	Adoption	Know-ledge	Adoption
1	Levelling of land	33.10	24.60	49.90	42.30	20.00	33.10
2	Contour bunding	26.90	25.30	33.07	26.12	40.00	48.40
3	Contour cultivation	40.00	61.50	36.90	12.30	23.10	26.11
4	Live bunding	22.30	12.30	21.50	26.20	56.10	61.50
5	Ploughing once in three years	45.30	43.80	41.50	31.50	13.07	24.61
6	Use of mulching	21.50	16.90	25.30	20.70	53.07	60.10
7	Crop planning on the basis of soil depth	23.07	23.07	27.60	36.90	49.20	40.30
8	Use of drought resistant varieties	23.07	24.60	53.80	35.31	23.07	40.00
9	Timely sowing	41.50	42.30	41.50	27.60	16.90	30.10
10	Strip cropping	37.60	33.08	36.10	31.50	26.10	40.30
11	Inter-cropping	36.10	31.50	28.40	22.30	35.30	46.11
12	Use of protective irrigations	32.30	24.60	42.30	33.07	25.30	42.33
13	Weed control	48.40	46.12	36.90	38.40	14.60	15.30
14	Optimum use of chemical fertilizers	31.50	19.20	40.70	20.70	27.60	60.10

Constraints faced by the dryland farmers in adoption of selected dryland agricultural innovations

The information on constraints faced by the dryland farmers in adoption of selected dryland agricultural innovations is given in Table 6.

Table 6: Distribution of Farmers according to Their Constraints in Adoption of Dryland Agricultural Innovations

Sr. No.	Constraints	Frequency	Percentage
A.	**Soil and water conservation practices**		
1.	**Levelling of land**		
a.	Lack of knowledge	26	20.00
b.	High cost	86	66.20
2.	**Contour bunding**		
a.	Lack of knowledge	52	40.00
b.	High cost	54	41.30
3.	**Contour cultivation**		
a.	Lack of knowledge	30	23.10
b.	Non-suitable and difficult to practice	33	25.38
4.	**Live bunding**		
a.	Lack of knowledge	73	56.10
b.	Non-availability of planting material	71	54.60
5.	**Ploughing once in three years**		
a.	Lack of knowledge	17	13.07
b.	High cost	73	56.15
6.	**Use of mulching**		
a.	Lack of knowledge	69	53.07
b.	Non-availability of mulching material	56	43.00
c.	High labour charges	74	56.90
7.	**Crop planning on the basis of soil depth**		
a.	Lack of knowledge	64	49.20
b.	Tradition of growing particular crop	74	56.90
B.	**General practices**		
8.	**Use of drought resistant varieties**		
a.	Lack of knowledge	30	23.07
b.	High cost	47	36.10
c.	Non-availability of seeds of varieties at the time of sowing	54	41.30
9.	**Timely sowing**		
a.	Lack of knowledge	22	16.90
b.	Uncertainty of rains	70	53.84
c.	Sowing delayed due to non-availability of bullocks	40	30.70
10.	**Strip cropping**		
a.	Lack of knowledge	34	26.10
11.	**Inter-cropping**		
a.	Lack of knowledge	46	35.30
b.	Non-availability of suitable seed drills	55	42.30
12.	**Use of protective irrigation**		
a.	Lack of knowledge	33	25.30
b.	Non-availability of water source	98	75.38

(*Continued*)

Sr. No.	Constraints	Frequency	Percentage
13.	**Weed control**		
a.	Lack of knowledge	19	14.60
b.	Lack of money to pay labour charges	66	50.76
c.	High cost of labour	42	32.30
14.	**Optimum use of chemical fertilizers**		
a.	Lack of knowledge	36	27.60
b.	High cost	89	68.40
c.	Uncertainty of rains	90	69.20
d.	Long drought spell	74	56.90

The major constraints reported by the farmers were lack of knowledge, lack of money, high cost of inputs, high labour charges, non-availability of planting material, lack of water sources for protective irrigations, etc.

Suggestions by the dryland farmers for overcoming the constraints in adoption of dryland agricultural innovations

The information regarding problem solution in order to overcome the constraints and to increase the levels of production and productivity in the dryland farming is given in Table 7.

Table 7: Suggestions of Dryland Farmers for Overcoming the Constraints

Sr. No.	Suggestions	Frequency N=130	Percentage Total 100%
I.	**About soil and water conservation**		
a.	Government should undertake and complete the work of land levelling	105	80.76
b.	Government should undertake and complete the work of contour bunding	88	80.76
c.	Financial assistance for digging of farm ponds or wells should be provided	94	67.00
II.	**About technical information on dryland farming**		
a.	Extension organization of state department of agriculture and agricultural universities should organize result demonstrations, field days, and training, etc. for transfer of improved dryland technology	93	72.30
b.	Extension literature on improved dryland farming technology should be prepared and supplied to the dryland farmers	95	71.50
c.	Climatic information related to dryland agriculture should be provided through mass media	89	73.70
d.	Farmers educational tours should be organized	108	68.40

Sr. No.	Suggestions	Frequency N=130	Percentage Total 100%
III.	**About inputs including credit**		
a.	Crop insurance coverage should be provided and benefits should be given immediately to the dryland farmers if crops fail	100	83.10
b.	Government should make loan available to dryland farmers at lower interest rate	103	76.92
c.	Government should provide critical inputs like seeds, fertilizers, farm implements to dryland farmers at reasonable prices	91	70.00

Conclusion

The majority of the dryland farmers were having partial to complete (full) adoption. There is a tremendous scope for increasing the adoption of dryland agricultural innovations on farmers field. This situation implies that extension organization of the state department of agriculture and SAUs should jointly plan and implement the transfer of technology programme for dryland agricultural innovations for the improvement of dryland farmers. The adoption of dryland agricultural innovations will surely increase if TOT programme is supported by the State Development Department and financial agencies like ZP, Banks and Insurance companies.

Effect of Soil Amendments on Improvement of Soil Physical Condition and Crop Growth of Pigeon Pea

3

G.J. Patel, D.B. Patel and *T.P. Kotecha*

Introduction

Pigeon pea is an important stable legume crop of command area. Two major soil types deep black and alluvial soil are found in the phase I of the Narmada command. The black soil are clayey in texture having very poor physical condition. The infiltration rate is low which causes run off, poor aeration and waterlogging and affect the crop production adversely with a view to improve the infiltration rate to the soil as well as crop growth the present study was undertaken.

Methodology

The field experiment consisted of seven treatments of soil amendments viz. control (T1), gypsums t/ha (T2), gypsum 10 t/ha (T3), gypsum 15 t/ha (T4), pressmud 5 t/ha (T5), pressmud 10 t/ha (T6) and pressmud 15 t/ha (T7). Pigeon pea (BDN-2) crop was sown during June with 8 kg/ha seed rate at an inter row spacing 90 cm × 20 cm. The experiment was laid out in randomized block design with four replications. The experimental site was clayey in texture, low in organic carbon (0.31%), medium in available K_2O (1012 kg/ha). The PH is 7.57 and EC is 1.23 dsm-1 in 1:25 ratio. All package of practices required for the crops were followed.

Result and Discussion

Grain and Haulm yield

The perusal of data in Tables 1 and 2 show that the significant difference were observed in grain yield and haulm yield of Pigeon

Table 1: Grain Yield (kg/ha) of Pigeon Pea as Influenced by Different Soil Amendments

Treatment	2000	2001	2002	Pooled
Control	1646	1569	1197	1471
5 t gypsum	1883	1850	1475	1736
10 t gypsum	1990	2097	1563	1883
15 t gypsum	2271	2211	1615	2032
5 t pressmud	2008	1792	1442	1747
10 t pressmud	1845	1963	1560	1789
15 t pressmud	2081	2014	1620	1905
C.D. at 50%	229	296	166	132
C.V.%	9.98	13.10	9.39	10.99

Table 2: Haulm Yield Pigeon Pea Influenced by Different Soil Amendments

Treatment	2000	2001	2002	Pooled
Control	2034	1883	1451	1789
5 t gypsum	2256	2214	1554	2006
10 t gypsum	2445	2506	1613	2188
15 t gypsum	2831	2652	1607	2363
5 t pressmud	2463	2160	1581	2069
10 t pressmud	2306	2356	1599	2088
15 t pressmud	2569	2419	1754	2247
C.D. at 50%	289	353	210	177
C.V.%	10.10	12.90	14.44	12.33

pea due to different soil amendments during all the three years and on pooled basis. More or less consistent results were observed during entire study. Application of 15 t/ha gypsum being at par with 15 t/ha pressmud gave significant higher grain yield than rest of treatments during 2000 and 2003. 15 t/ha gypsum being at par with 10 t/ha gypsum and 15 and 10 t/ha pressmud established its significant superiority during 2001. On pooled basis, the treatment of application of 15 t/ha gypsum and pressmud gave almost equal were significantly superior that rest of the treatments. In general application of pressmud or gypsum @ 15 t/ha increase the grain yield 38 and 30 per cent respectively over without soil amendments. The results on haulm yield revealed similar trade to grain yield. Application of either pressmud or gypsum ranging from 5 to 15 t/ha increased the haulm yield to the tune of 20 to 37 per cent.

Soil physical condition

Data presented in Table 3 show that soil amendments, i.e. application of pressmud and gypsum improve the soil properties, organic carbon, infiltration rate and water holding capacity. In respect to the organic carbon content of the soil was higher in the treated plots. Application of 15 t/ha pressmud recorded higher values.

Table 3: Effect of Organic Amendments on Soil pH, EC (d Sm^{-1}), O.C. (%), Infiltration Rate (cm/hr) and WHC (%)

Treatment	PH (1.25)	EC (d Sm^{-1})	O.C. (%)	Bulk density	Infiltration rate (cm/hr)	WHC (%)
Control	7.558	1.60	0.345	1.295	1.105	58.775
5 t gypsum	7.368	1.35	0.420	1.295	1.590	59.225
10 t gypsum	7.395	1.30	0.420	1.295	1.710	60.025
15 t gypsum	7.322	1.05	0.485	1.295	1.728	61.050
5 t pressmud	7.535	1.33	0.457	1.287	1.795	60.800
10 t pressmud	7.560	0.93	0.485	1.287	1.867	62.275
15 t pressmud	7.585	0.78	0.540	1.285	1.920	62.875
S.Em+	0.04	0.22	0.02	0.01	0.05	0.24
C.D. 5%	0.13	NS	0.06	—	0.14	0.72
C.V.%	1.20	36.95	9.54	0.80	0.80	5.79

Infiltration rate was improved with the application of soil amendments viz., gypsum, pressmud, highest being under the treatment of 15 t/ha pressmud which was about twice than the control. Similar observations were made by Bharambe et al. (1990).

Water holding capacity (WHC) was higher following the application of pressmud at 15 t/ha which was at par with pressmud at 10 t/ha, gypsum 15 t/ha. The considerable reduction in pH and EC showed that soil amendments studied, were found effective in the clay soil in Narmada command area under pigeon pea cropping system.

Economics

The data presented in Table 4 revealed that the application of 15 t/ha gypsum realized the maximum net realization of Rs. 22,988 ha^{-1} which was followed by 10 t/ha gypsum and

Table 4: Economics of Treatment

Treatments	Grain yield (kg/ha)	Haulm yield (kg/ha)	Gross realization (Rs/ha)	Cost of cultivation (Rs/ha)	Net realization (Rs/ha)
Control	1471	1789	26538	9875	16663
5 t gypsum	1709	2006	30650	11125	19525
10 t gypsum	1883	2188	33715	12375	21340
15 t gypsum	2047	2363	36613	13625	22988
5 t pressmud	1747	2069	31378	10875	20503
10 t pressmud	1789	2088	32055	11875	20180
15 t pressmud	1905	2247	34193	12875	21318

Selling price of pigeon pea 15.0 Rs/kg Selling of haulm 2.5 Rs/kg

Cost of gypsum Rs. 250 per tonne Cost of pressmud Rs. 200 per tonne.

15 t/ha pressmud which gave Rs. 21,340 and Rs. 21,318 ha^{-1} respectively in comparison to control and rest of the treatments.

From the foregoing results it becomes evident that soil amendments like gypsum and pressmud improving the physical characters like infiltration rate and WHC of clay soil increase grain yield as well as monitory net realization.

REFERENCE

Bharambe, P.R., Awasarmal, B.C., Ambegaonkar, P.R. and Rodge, R.P. (1990). *J. Indian Soc. Soil Sci.*, 38 (3): 488-94.

Decision Making Pattern of Dryland Farmers in Tamil Nadu

4

K. Thangaraja, C. Karthikeyan and *J. Venkatapirabu*

Introduction

In Tamil Nadu, area under dry farming constitutes 52 per cent of total cultivable area contributing to 40 per cent of total food production. The productivity of crops grown in dryland is not only low but also remain stagnant over years. Most of the areas in Tamil Nadu come under semi-arid tropical climate, except the hilly regions and east coast. Out of 5.50 m ha of net sown area, nearly 3.20 m ha are rainfed (Kannaiyan et al., 2001). Majority of the dryland farmers are small farmers with scarce resources. The poor resource base allows only a low input subsistence farming with low and unstable crop yield.

Decision making is important because much of the success of any enterprise and particularly farming depends upon how well the family makes decision. Every action of individual is the result of conscious or unconscious decision. The present decision has its roots in the past and reflects upon the future decision. The decision is the heart of farm management. To know about dryland farmers in decision making pattern of various agricultural activities in decision and other enterprises it becomes imperative to understand their willingness first about self decision making and then deciding after consulting the spouse/elders in the family and consulting others (neighbours, progressive farmers, relatives and extension officers) in decision making.

For the purpose of studying the decision making pattern in dryland farmers, the major decision making areas were identified and classified into nine major headings like purchase of agricultural inputs, getting loans/credits, investment decision, seeds and sowing, inter-cultivation, harvesting, quantity of produce to be stored, allied activities and marketing decisions. In this context, the present study was taken up with the specific objective viz., to study the decision making pattern of dryland farmers in agricultural and allied activities of dry farming.

Methodology

The study was conducted in three revenue villages of Reddiarchatram block of Dindigul district in Tamil Nadu. The sample size was 90 farmers consisting of 45 maize growers and 45 sorghum growers. 30 farmers were drawn randomly from each village. Thus, 10 farmers were selected from each of the nine villages covering three revenue villages using simple random sampling technique. The data were collected with well structured and pre-tested interview schedule. Percentage analysis was carried out for getting meaningful interpretation of results. The major findings of the study are as follows.

Findings and Discussion

Decision making pattern of dryland farmers

In farming families not only individual decisions were taken by the head of the family, but also joint decisions were taken by consulting spouse, elders in the family and consulting others (namely neighbours, progressive farmers, relatives and extension officers). The decision making pattern of farm families and dryland farmers was assessed for agricultural and other enterprises in their farm. The results are presented in Table 1.

Purchase of agricultural inputs

It could be understood from Table 1 that changes had occurred in decision making pattern of dryland farmers. Regarding purchase of agricultural inputs like seeds 37.78 per cent of the farmers had taken self decision followed by joint decisions consulting spouse/elders in the family (20%) and consulting others (8.89% neighbours, 3.33% progressive farmers, 12.22% relatives, 17.78% extension officers respectively). Similarly, fertilizers 43.33 per cent of the farmers had taken self decision followed by joint decision (22.22%) and consulting others (5.56% neighbours, 10.0% progressive farmers, 8.89% relatives, 10% extension officers respectively) with regard to the purchase of plant protection chemicals 31.11 per cent of the farmers had taken self decision followed by joint decision (16.67%) and consulting others (8.89% neighbours, 13.33% progressive farmers, 11.11% relatives, 18.89% extension officers respectively).

Getting loans/credits

Regarding getting loans/credit, more than half of the farmers (51.11%) had taken self decision followed by joint decision (27.78%) and consulting others (4.44% neighbours, 8.89% relatives and 7.78% extension officers respectively).

Investment decision

About investment decision like buying/selling/leasing of the land individual decision was taken by majority of the farmers (60%), 20 per cent of the farmers reported taking decision by consulting the spouse/elders in the family. The rest of farmers took decision by consulting others (6.67% neighbours, 13.33% relatives). Hiring of equipments, individual decision was taken by 40 per cent of the farmers, 25.56 per cent of the farmers reported for decision by consulting the spouse/elders in the family. The remaining (5.56% neighbours, 3.33% relatives, 16.67% extension officers respectively) farmers had taken decision by consulting others with regard to engaging labourers majority of the farmers (71.11%) had taken decision by themselves followed by joint decision through consulting the spouse/elders in the family (28.89%).

Seeds and sowing

Regarding seeds and sowing like using recommended seed rate 45.56 per cent of the farmers had taken self decision followed by joint decision by consulting spouse/elders in the family (16.67%) and consulting others (7.78% neighbours, 4.44% progressive farmers, 10% relatives, 15.56% extension officers respectively). As regards the sowing season and time more than half of the farmers (52.22%) had taken individual decision followed by one-fourth of the farmers (25.56%) had taken joint decision and by consulting others (5.56% neighbours, 3.33% progressive farmers, 4.44% relatives and 8.89% extension officers respectively in that order. Likewise in the case of method of sowing a majority of the farmers (86.67%) reported self decision followed by 13.33 per cent of the farmers took decision by consulting the spouse/elders in the family.

Intercultivation

About intercultivation operations like weeding, a majority of the farmers (75.56%) had taken self decision followed by joint decision (24.44%). With regard to the type of fertilizers, nearly half of the farmers (47.78%) had taken self decision followed by joint decision (18.89%) and consulting others (7.78% neighbours, 10% relatives and 15.56% extension officers). In the case of time of fertilizer application, 57.78 per cent of the farmers had taken self decision followed by one-fourth of the farmers had taken joint decision (24.44%) and consulting others (6.67% progressive farmers, 11.11% extension officers). As regards the time and stage of pesticide application, 25.55 per cent of farmers had taken self decision followed by joint decision (17.78%) and consulting others (15.56% neighbours, 13.33% progressive farmers, 7.78% relatives and 20% extension officers).

Harvesting and storage

About time of harvesting majority of the farmers (68.89%) had taken self decision followed by joint decision (31.11%). With regard to the method of harvesting cent per cent of the farmers had taken self decision only. Regarding quantity of produce to be stored 46.67 per cent of the farmers had taken self decision followed by joint decision (25.56%) and consulting others 13.33% (neighbours 5.56%, progressive farmers 3.33% and 8.89% relatives).

Allied activities

Regarding the number of animals to be maintained in a dairy farm, individual decision was taken by majority of the farmers (62.22%). About 37.78 per cent of the farmers reported taking decision by consulting the spouse/elders in the family. As regards the disease treatment (dairy, poultry) 13.33 per cent of the farmers had taken self decision followed by joint decision

(20%) and consulting others (11.11% neighbours, 13.33% progressive farmers, 7.78% relatives and 34.44% extension officers). In the case of feeding time and quantity, a majority of the farmers (62.22%) had taken self decision followed by joint decision (37.78%).

Marketing decisions

It could be observed from Table 1 that regarding marketing of agricultural produce, 41.11 per cent of the farmers had taken individual decision followed by joint decision (16.67%) and consulting others (10% neighbours, 15.56% progressive farmers and 13.33% relatives). In the case of marketing of dairy products 43.33 per cent of the farmers had taken self decision followed by joint decision (23.33%) and consulting others (16.67% neighbours, 6.67% progressive farmers and 10% relatives). Likewise in the case of poultry marketing, 43.33 per cent of the farmers had taken self decision followed by joint decision (17.78%) and consulting others (13.33% neighbours, 8.89% progressive farmers and 16.67% relatives).

Regarding quantity of produce to be sold, majority of the farmers (71.11%) had taken self decision followed by joint decision consulting spouse/elders in the family (28.89%). As regards the selling time, individual decision was taken by 57.78 per cent of the farmers, 16.67 per cent of the farmers reported for taking decision by consulting the spouse/elders in the family. The rest of the farmers took decision by consulting others (8.89% neighbours, 5.56% progressive farmers and 11.11% relatives). Regarding selection of the place to market the produce half of the farmers (51.11%) had taken individual decision followed by joint decisions consulting spouse/elders in the family (20%) and consulting others (6.67% neighbours, 8.89% progressive farmers and 13.33% relatives).

Conclusion

From this study it could be concluded that the dryland farmers had resorted to all methods of decision making and majority of them took decision individually about agricultural and allied activities. Self decision makers were more in the case of purchase of agricultural chemicals, getting loans/credits, investment decision like buying/selling/leasing of land, hiring of equipments and engaging labours, seeds and sowing use recommended seed rate, sowing season and time, method of sowing, inter-cultivation operation like weeding, type of fertilizers, time of fertilizer application, time of harvesting and method of harvesting, quantity of produce to be stored, marketing about crop, dairy product and poultry, quantity of produce to be sold, selling time and selection of market place followed by only few joint decision in the case of management of a dairy farm.

REFERENCE

Kannaiyan, S., Thiyagarajan, T.M., Subramanian, M., Balasubramanian, T.N. and Selvaraj, R. (2001). *Dryland Green Revolution in Tamil Nadu: The Perspectives*. Coimbatore: Tamil Nadu Agricultural University Press, pp. 10-12.

Table 1: Decision Making Pattern of Dryland Farmers (n = 90)

S.No.	Decision making areas	By self without consulting others		Consulting the spouse or elders in the family		Consulting others							
						Neighbours		Progressive farmers		Relatives		Extension officers	
		No.	%	No.	%	No.	%	No.	%	No.	%	No.	%
1	2	3	4	5	6	7	8	9	10	11	12	13	14
1.	Purchase of agricultural inputs												
(a)	Seeds	34	37.78	18	20.00	8	8.89	3	3.33	11	12.22	16	17.78
(b)	Fertilizers	39	43.33	20	22.22	5	5.56	9	10.0	8	8.89	9	10.0
(c)	Plant production chemicals	28	31.10	15	16.67	8	8.89	12	13.33	10	11.11	17	18.89
2.	Getting loans/credits	46	51.11	25	27.78	4	4.44	–	–	8	8.89	7	7.78
3.	Investment decision	–	–	–	–	–	–	–	–	–	–	–	20.00
(a)	Buying/selling/leasing of land	54	60.00	18	20.00	6	6.67	–	–	12	13.33	–	–
(b)	Hiring of equipments	36	40.00	23	25.56	5	5.56	3	3.33	8	8.89	15	16.67
(c)	Engaging labours	64	71.11	26	28.89	–	–	–	–	–	–	–	–
4.	Seeds and sowing												
(a)	Seed rate	41	45.56	15	16.67	7	7.78	4	4.44	9	10.00	14	15.56
(b)	Sowing season and time	47	52.22	23	25.56	5	5.56	3	3.33	4	4.44	8	8.89
(c)	Method of sowing	78	86.67	12	13.33	–	–	–	–	–	–	–	–
5.	Intercultivation	–	–	–	–	–	–	–	–	–	–	–	–
(a)	Weeding	68	75.56	22	24.44	–	–	–	–	–	–	–	–

Table 1 (Contd.)

1	2	3	4	5	6	7	8	9	10	11	12	13	14
(b)	Type of fertilizer	43	47.78	17	18.89	7	7.78	–	–	9	10.00	14	15.56
(c)	Time of fertilizer application	52	57.78	22	24.44	–	–	6	6.67	–	–	10	11.11
(d)	Time of stage of pesticide application	23	25.55	16	17.78	14	15.56	12	13.33	7	7.78	18	20.00
6.	Harvesting												
(a)	Time of harvesting	62	68.89	28	31.11	–	–	–	–	–	–	–	–
(b)	Method of harvesting	90	100.00	–	–	–	–	–	–	–	–	–	–
7.	Quantity of produce to be stored	42	46.67	23	25.56	12	13.33	5	5.56	8	8.89	–	–
8.	Allied activities												
(a)	Number to be maintained	56	62.22	34	37.78	–	–	–	–	–	–	–	–
(b)	Treatment to be taken	12	13.33	18	20.00	10	11.11	12	13.33	7	7.78	31	34.44
(c)	Feeding time and quantity	56	62.22	34	37.78	–	–	–	–	–	–	–	–
9.	Marketing												
(a)	Crop	37	41.11	18	20.00	9	10.00	14	15.56	12	13.33	–	–
(b)	Dairy	39	43.33	21	23.33	15	16.67	6	6.67	9	10.00	–	–
(c)	Poultry	39	43.33	16	17.78	12	13.33	8	8.89	15	16.67	–	–
i.	Quantity to be sold	64	71.11	26	28.88	–	–	–	–	–	–	–	–
ii.	Selling time	52	57.78	15	16.67	8	8.89	5	5.56	10	11.11	–	–
iii.	Selection of market place	46	51.11	18	20.00	6	6.67	8	8.89	12	13.33	–	–

Appropriate Farming and Extension Strategies for Wasteland and Dryland Areas: An Empirical Matrix

5

K. Chandrakandan, C. Cinthia Fernandaz
and *R. Ganesan*

Introduction

We have 142 m ha cultivable area in which one-third is under irrigation, remaining is dryland. In dryland areas crop failure is a common phenomenon faced by the farmers. A few of them have resorted to some allied enterprises to invert risks and get additional income and employment opportunity from per unit area. With the feasibility of integration of various enterprises, this paper is to analyse an appropriate farming and extension strategies for waste and drylands. The variables such as education, decision making, farm size, annual income, social participation, economic motivation and level of aspiration significantly differed between the practitioners and non-practitioners of Integrated Farming System (IFS).

Perception on modern IFS technologies differed significantly between the two groups, when all IFS were considered together, i.e. combined data. Almost all the respondents (98.33%) belonged to high perception group, whereas majority (97.5%) of the exclusive crop growers come under medium perception group only. The variables economic motivation and farm power had significant relationship with perception in the crop alone system. In the case of IFS, farm size, annual income and farm power were found to be significantly related with perception. Contribution of farm power was more in exclusive crop farming system. In IFS knowledge level and education had significant contribution towards perception.

The 't' test showed the significant difference between the IFS practicing and non-practicing farmers with respect to their training need. The training needs on general aspect of IFS were integration of various enterprises in the farm, management of IFS enterprises, type and size of enterprises (to be included) in that order of importance.

In crop alone system, level of aspiration had significant relationship with training needs and also contributed much towards training needs, whereas in IFS, social participation and level of aspiration had significant relationship with training needs and social participation showed highly significant contribution towards the training needs.

The Cost: Benefit ratio was high for agri-horticulture, i.e. 2.34 whereas it was 1.73 in the case of silvipasture, 1.68 in the case of sericulture, 1.59 in the case of apiculture, 1.45 in dairy and 1.05 in poultry.

Appropriate Implications

- There is a need for creating mass awareness among non-IFS practitioners and impressing them to increase their perception level and thereby adopt IFS practices.
- An integrated and need based training should be planned and organised.
- ITK can be used for refinement of technology.
- As the study revealed a vast scope and feasibility for popularising of IFS, more drive and extension efforts may be made among, non-adopted areas and farmers.

Appropriate IFS Matrix

Based on the findings, the following appropriate IFS matrix is suggested:

1. Resource poor farmers
   - → Marginal (up to 2.5 ac) → Agri-horticulture, Poultry, Apiculture, Sericulture
   - → Small (2.5 to 5.0 ac) → Agri-horticulture, Dairy, Silvipasture, Apiculture
2. Resource rich farmers
   - → Medium (5 to 10 ac) → Poultry, Sericulture
   - → Large (above 10 ac) → Agri-horticulture, Poultry, Silvipasture, Apiculture

Resource Poor Farmers

Marginal farmers

Marginal farmers having alfisol soil type (red soil) and cultivating crops with normal rainfall may include agri-horticulture, horticulture, one or two dairy animal, backyard poultry (100 birds), apiculture and sheep/goat (3-4 numbers) in their farms.

Farmers having vertisol soil (black soil) can adopt small poultry units, apiculture, sheep and goats along with cropping system in their farms. Farmers having irrigation facilities in vertisol soils may include sericulture (0.2-0.4 ha), agri-horticulture and dairy to get additional income from the same unit of land. Sericulture is labour intensive and hence more suitable to marginal and small farmers having family labour to take care of rearing worms themselves.

Small farmers

In alfisol soil (red soil) belt, farmers can grow crops based on monsoonic rains availability and they can have agri-horticulture, dairy (2 to 4) animals, apiculture and silvipasture along with cropping. Those who have little irrigation facility even in the rainfed area can include horticulture crops, dairy, sericulture (0.4 to 0.8 ha) silvipasture and other enterprises.

Farmers of vertisol soil (black soil) belt, having only rain fed area may adopt agri-horticulture (even rainfed vegetables, banana, etc.), dairy (3 to 4)/animals apiculture, silvipasture and sheep and goat rearing. If they have irrigated as well as rainfed areas means, they can have horticulture crops mostly vegetables, sericulture (0.4 to 1 ha) and some other enterprises. Farmers having land nearby forest means, they can adopt agriculture, farm forestry and the like.

Resource Rich Farmers

Medium farmers (Medium size holdings)

In this category, farmers may be able to spend adequate money either own or borrowed for introducing enterprises in their farms. In alfisol soil, rainfed land holders may include

Constraints Faced by the Respondents

Sl. No.	Constraints	Number	Percentage
	On IFS (n=120)		
1.	Lack of inputs availability	54	45.00
2.	High labour cost	65	54.16
3.	Lack of rainfall	115	95.83
4.	Lack of technical assistance	108	90.00
5.	Lack of credit facilities	85	70.83
	On crop production (n=160)		
1.	Lack of inputs availability	75	45.00
2.	High labour cost	101	63.12
3.	Poor/lack of extension service	139	86.87
4.	Uncertainty in the price of the produce	91	56.87
5.	Lack of rainfall	133	83.12
	On dairy farming (n=30)		
1.	Unavailability of green fodder in lean season	25	83.33
2.	Inadequate veterinary services	20	66.66
3.	Lack of availability of fodder	30	100.00
4.	High cost of concentrates	28	93.33
5.	Lack of knowledge on fodder crop cultivation	20	66.66
	On sericulture (n=30)		
1.	Fluctuation in the price of cocoon	18	60.00
2.	High cost of egg cards	21	70.00
3.	Lack of disease control facilities	29	96.66
4.	High risk	24	80.00
	On poultry (n=30)		
1.	High cost of poultry feed	30	100.00
2.	Fluctuation in the price of eggs	30	100.00
3.	Lack of training on poultry health management	29	96.66
4.	Lack of technical assistance	14	46.66
5.	High risk	17	56.66
	On agri-horticulture (n=30)		
1.	More pest and disease attack	30	100.00
2.	Lack of technical assistance	20	66.66
3.	Lack of rainfall	30	100.00
4.	High cost of plant protection chemical	15	50.00

poultry, sericulture, etc. depending on family labour. But farmers having irrigation facilities or water resource may include all enterprises except apiculture if they are not having lands nearby forest area for honey bee activity.

Large farmers (Large holdings)

In alfisol area, rainfed farmers can adopt poultry (on commercial scale), agri-horticulture, silvipasture, apiculture and sheep/goat farm along with their cropping system. Some farmers may have borewell or open well for irrigation purpose. They can adopt horticulture and poultry with other enterprises and crop farming.

Based on research evidence and observations we believe that different categories of farmers can adopt varied enterprises as suggested above, depending on the farmer's resources, land type, rainfall pattern and irrigation supplements available and their innovativeness. Progressive farmers have already shown the new ways and so the extension system may make use of them as "model farms" for popularising such new ventures among the rest of the farming community.

Farmers reported five major constraints in practicing IFS. The most important constraint reported to have been faced by 95.83 per cent of the respondents was lack of rainfall. In dry lands and particularly in semi-arid zone, failure of rainfall is a major problem. Due to monsoon failure, farmers are not able to sow crops in season. Pre-monsoon sowing and release of drought tolerant varieties which are suitable to all seasons may be the remedy to this problem. Effective cropping system by using available rainfall should be introduced to offset the crop failure.

The second important constraint faced by 90 per cent of the respondents was lack of technical assistance. In IFS, farmers are practicing many enterprises, like dairy, poultry, sericulture, agri-horticulture, etc. There are separate departments for agriculture, horticulture, animal husbandry and sericulture. Farmers face difficulties to get technical assistance from these departments. They wanted single window system to get whatever they want. Village level workers (VLWs) are not having the capacity to answer all the questions raised by the farmers. The capacity of the VLWs should be improved by regular training and their jurisdiction also reduced to contact maximum number of farm families instead of selected ones. The model of Broad Based Extension System will be helpful to farmers for getting technical assistance.

The third important constraint was lack of credit facilities (70.83%). Most of the farmers belonged to marginal and small farmers categories. IFS requires more funds for starting any mixed enterprises. They are not in a position to start with their own money. They prefer to borrow from village co-operative societies. Societies and banks refused to lend credit for more than one enterprise at a time for the farmers. For credit availability to the farmers, societies should give long-term loan on the basis of their enterprises selection with technical guidance.

High cost of labour was felt as the next important constraint by 54.16 per cent of the respondents. IFS is a labour intensive practice. Skilled labourers are required to do special operations. For this, farmers are forced to pay more. By increasing the income of a farmer, he cannot face high labour cost. Due to getting high cost-benefit ratio, farm mechanisation also will give remedy to this problem.

Lack of inputs availability was faced by 45 per cent of the farmers. Normally farmers bought inputs from agro-service centres by spending more money. Even though it is high priced, inputs are not given with good quality. So farmers are not able to get adequate inputs. Government or co-operatives should supply required quality and quantity of inputs through its depots by single window system, IFS farmers can get inputs, credit, advice, marketing facilities, information, etc.

Extension Approach/Strategy to be Adopted for Introducing IFS among Non-practicing Farmers

Individual, group and mass approaches are normally used by extension personnel to create awareness of new technology or programmes among the farming community.

Among various approaches, group meeting, campaigns, training, demonstration, filed trips/study tour and audio-visual aids can be used for all types of farmers. Field trips to model farms will give first hand information about IFS and related aspects. It may be useful to get adequate knowledge through audio-visual aids.

Demonstration helps to learn skills or how to make use of the knowledge for adopting a particular enterprise. Result demonstrations may be very effective for convincing and audio-visual aids can motivate farmers to adopt different enterprises in their farms.

Training should be given to the farmers to get adequate knowledge and skills for adopting enterprises like sericulture, apiculture, dairy, etc. and also for better management, more so when the new technologies are introduced.

In this world of ICT, resourceful farmers may have contact through Internet to the university scientists and other research stations to get the latest information. Cyber extension system may be used to medium and large farmers who are more educated and resourceful. In this, computer aided digital and CD may be used for getting information about any enterprise. Instead of disseminating general information from the research centers, specific information needed by the farmers may induce farmers for adopting appropriate IFS. After that, mass media like television, newspaper, radio, etc. can be used to update the knowledge level of farmers and further reinforcement. To sum up, it is suggested that all the development departments may provide right messages in a right manner, to the right people in right season in a coordinated way to trigger the desired action to achieve desirable outcomes and behavioural changes.

REFERENCES

Iqbal, Mohamed Ismail (1992). *Integrated Farming System: Research, Extension and Clientele Analysis*. Unpub. M.Sc (Ag.) Thesis, TNAU, Coimbatore-3.

Lakshmi, T. and M. Manoharan (1994). Knowledge on Dryland Technology. J. Extn., 5(2): 858-65.

Mary, Elizabeth (2001). *Integrated Dry Farming System in Tamil Nadu—A Feasibility Study*. Unpub. MSc (Ag.) Thesis, TNAU, Coimbatore-3.

Pushpa, J. (1996). *Impact of Intergrated Farming Systems on Garden Land Farmers*. Unpub. Ph.D. Thesis, TNAU, Coimbatore-3.

Wilson, W. (2001). *The Fulani Model of Sustainable Agriculture*. Unpub. Manuscript. African American Studies, University of Marvland, Baltimore.

Scope of Refining and Promoting ITKs on Soil and Water Conservation in Rainfed Areas

6

P.K. Mishra, M. Osman and *G.R. Maruthi Sanker*

Introduction

Soil and water are the basic resources and these must be conserved as carefully as possible. The pressure of increasing population neutralises all efforts to raise the standard of living, while loss of fertility in the soil itself nullifies the value of any improvements made. This calls for more systematic resource conservation efforts. It is well-known to every farmer that it is the top soil layer, which sustains agricultural production. Once this layer is lost or eroded, nothing can be done to replace it within a short period of time. Climate and hydrology, soil topography, soil surface conditions and their interactions are major factors affecting erosion-sedimentation processes. The semi-arid regions with few intense rainfall events and poor soil cover condition produce more sediment per unit area. But the man's intervention has disturbed the natural equilibrium and intensive and extensive agriculture has become a dominant factor in accelerating land degradation. The ever-increasing population pressure has brought intensive cultivation of land to the forefront through irrigated agriculture. No doubt, these practices have resulted in a great increase in productivity, but they have resulted in large-scale waterlogging. Cultivable wastelands are increasing in the agricultural fields due to improper land management. The obvious remedy for this is to follow soil and moisture conservation practices along with integrated nutrient supply system for improvement of soil fertility as well as crop productivity on sustained basis. Soil conservation in any form is the only known way to protect the productive lands. In a predominantly agricultural country like India, where droughts and floods cause chronic food scarcity, adequate soil conservation programme, not only increases crop yield, it also prevents further deterioration of land. Methods to control surface run-off and soil associated erosion have been practiced in India from times immemorial.

It is estimated that it takes 400 to 1000 years for formation of just 2.5 cm of topsoil. Left to nature, the soil is protected by natural vegetation. Because of human interference, the protective shield of land is disturbed and the land is torn into pieces by the erosion process—starting from splash erosion to formation of gullies and ravines. Man's ever increasing needs resulted in felling of trees and removal of grasses. As a result, erosion problem has assumed threatening dimensions. Realising this, Government of India has given top priority for treating rainfed areas with intensive soil and water conservation measures for controlling erosion as well as harvesting rainwater. For getting greater success and better dividends, the rainwater management should be carried out on watershed basis.

The term Indigenous Technical Knowledge (ITK), "local knowledge" and "traditional knowledge" have been used in the literature inter-changeably. Traditional knowledge is gathered over a period of time and transferred from generation to generation. It is synonymous to local knowledge and is defined as "A sum total of knowledge based on acquired knowledge and experience of people in dealing with problems and typical situation in different walks of life." It is the knowledge which has been accumulated by the people over generations by observation, by experimentation and by handling on old peoples' experiences and wisdom in any particular area of human behaviour. Indigenous technical knowledge is the local knowledge that people have gained through inheritance from their ancestors. It is a people derived science and represents people's creativity, innovations and skills. Indigenous technological knowledge pertains to various cultural norms, social roles or physical conditions. Such knowledge is not a static body of wisdom, but instead consists of dynamic insights and techniques, which are changed over time through experimentation and adoption to environmental and socio-economic changes. This knowledge has backgrounds of hundreds and sometimes thousands of years of adoption while bearing odds and evens of the time.

Fortunately, we have many indigenous techniques for conserving natural resources (Agarwal and Narain, 1999). These have been in practice for number of years as presented in the write up. Therefore, there is a need to enmesh these practices along with conventional soil and water conservation measures for promoting sustainable development of agriculture. It may not be out of place to mention that some of these ITKs may need minor modifications in different watershed situations as well as socio-economic fabrics across the country. Inclusion of these ITKs would ensure sustainability of different eco-systems, befitting the man-animal-plant-land-water complex in each watershed. The documentation of ITKs on soil and water conservation will form a basis for formulating coordinated research programme for validation and refinement of the ITKs on soil and water conservation. In recent years the idea of taking ITK into consideration in developing projects and locating research thrust areas has been gaining momentum. This knowledge is not possessed by only one sector of society, for example, in many cultures, women and elders have passive insights into certain aspects of culture. Sometimes, researchers have been unaware of such perceptiveness among rural people due to their biased focus on land-owning male farmers, neglecting other members of society. Traditional knowledge and practices have their own importance as they have stood the test of time and have proved to be efficacious to the local people. Some of these traditional practices are in the fields of agriculture such as crop production, mixed farming, water harvesting, conservation of forage, combined production system, biodiversity conservation, forestry and domestic energy, etc. India is unique having a rich history of traditional systems of soil conservation and water harvesting in almost all the states. Conservation of both surface and groundwater has been an integral part of our country for many centuries. In fact, different types of ponds and tanks represent important community resources for drinking water and allied uses in rural India. Even today, the main attributes to their success are the sound scientific knowledge and methods on which they have been built. Moisture conservation begins right from seedbed preparation. Although farmers practice many indigenous technologies relating to soil and water conservation, there is a lack of documentation for identifying the constraints for possible refinements. There exists a need to evaluate the potential indigenous practices in the regions for their improvement and dissemination to new areas. There is also a need for scientist-farmer interaction for large-scale adoption.

Methodology

In order to obtain the feedback of the farmers regarding soil and water conservation measures the survey using developed proforma was initiated in treated and untreated villages in different agro-ecological regions through the project centres. Farmers' awareness and adoption of soil and water conservation practices was attempted in the study. The study was carried out in 18 centres of All India Coordinated Research Project on Dryland Agriculture (AICRPDA) including Hyderabad, which was the nodal centre of the project. These centres cut across various states and various rainfall zones but all represent rainfed zones of the country. The process followed during documentation was as follows:

1. The ITK proformae consisting of general information of village, views and observations on soil and water conservation measures, *in situ* moisture conservation practices, technical information on run-off management practices and indigenous technical knowledge was developed.
2. The proformae for documenting ITKs on soil and water conservation and run-off management was finalised duly after getting the feedback from participants.

For the purpose of documentation and analysis of ITK, a simple ITK proforma was developed. This proforma contains Title, Purpose, Location, Agro-ecological setting, Description, Advantages, Constraints, Replicability/Feasibility and

Researchable issues. The researchable issue opens up new vistas for furthering the technologies towards development of farmer-friendly doable technologies. This is just an honest beginning of a validation process made by NATP, ICAR. Of the many available, a few ITKs could be documented in the limited time-frame through the project centres located in various SAUs. There is a lot more to be done in this direction by the R&D and implementing agencies involved in the community development.

Documented ITKs

The ITKs were documented under following specific categories (Table 1):

- Agronomic Measures
- Tillage
- Bunding and Terracing (Mechanical and Vegetative barrier)
- Land Configuration
- Soil Amendment/Mulching
- Erosion Control and Run-off Diversion Structures
- Water Harvesting, Seepage Control and Groundwater Recharge.

Table 1: A list of Some Documented ITKs on Soil and Water Conservation Measures under Different Categories

Sl.No.	Categories	Name of ITK
1	Agronomic Measures	Inter-cropping
		Cultivation and sowing across the slope
		Wider row spacing and deep inter-culturing
		Mixed cropping
		Cover cropping
		Criss-cross ploughing
		Hoeing with local hoes
		Set furrow cultivation
		Application of manure (FYM)
		Strip cropping
		Green capping
		Green manuring
		Pre-emergence soil stirring
		Ridge and furrow planting

Sl.No.	Categories	Name of ITK
2	Tillage	Conservation furrows with traditional plough Deep ploughing Summer ploughing/Off-season tillage Repeated tillage during monsoon season
3	Bunding and Terracing (Mechanical and Vegetative barrier)	Vegetative barrier Stone bunding *Nala* check with soil filled in cement bags Compartmental bunding Peripheral bunding/Field bunding Ipomoea as vegetative barrier Conservation bench terrace Loose stone surplus Stabilization of field boundary bund with *Vitex negundo* Strengthening bunds by growing grasses Bund farming of pulse crops in *kharif* under rainfed situation Earthen bunds Stone-cum-earthen bunding Live bunding by raising Cactus Grass plantation on field boundaries (filter strip) Growing of Saccharum munja as vegetative barrier on field boundaries
4	Land Configuration	Use of indigenous plough for formation of broad bed and furrows Furrow opening in standing crops local implement hoe (Dawara) for moisture conservation Levelling the plots by local leveler Opening up set furrow Conservation furrow: *Gurr*
5	Soil Amendment/ Mulching	Application of tank silt Application of groundnut shells Sand mulching Gravel sand mulching Retention of pebbles on the soil surface Retention of sunflower stalks Mulching of *Sal* leaf in turmeric Crop residue application in the field
6	Erosion Control and Run-off Diversion Structures	Sand bags as gully check Loose boulder checks Stone waste weir Waste weir (stone/sorghum stubbles) at the outlet of the field Brushwood structure across the bund Grassed waterways

(*Continued*)

Sl.No.	Categories	Name of ITK
		Spur structure *Nala* plugging
7	Water Harvesting, Seepage Control and Groundwater Recharge	Seepage control by lining farm ponds with white soil Harvesting of seepage water Wells as run-off storage structures Rainwater management using indigenous rain gauge (*Role*) Farm pond Percolation pond/tank Groundwater recharging through ditches and percolation pits Well recharging through run-off collection pits Dug wells Haveli/Bharel system *Bandh* system of cultivation Earthen check dams Field water harvesting *Nadi* farming system Collection of sub-surface run-off water and recycling in *Diara* land Rainwater harvesting from roof top and road surfaces Rainwater harvesting in *Kund/Tanka*

Refinement of ITKs for promotion of the technologies

Some potential ITKs identified for further study, research and development of new projects is presented in Table 2. A scientific study may change this Indigenous Technical Knowledge to Modern Technical Knowledge (MTK). Prevailing ITKs should invariably be given priority. All the ongoing projects on resource conservation and management should focus on the viable and appropriate ITKs relating to soil and water for sustainable development and dissemination of the local technology.

As an initiative, the ITKs on Soil and Water Conservation (S&WC) for other agro-climatic regions may be documented and later validated and refined at local level. This will form a programme by itself to popularise indigenous knowledge with the developmental agencies. Exposure visit and farmer-to-farmer interaction on the subject may be encouraged for better adoption. The ongoing watershed programme should adopt the ITKs on S&WC in their project activities. During the first phase planning the local technologies should be documented.

The stakeholders in the conservation programme who can be partners for promoting the ITKs are:

1. Farmers
2. NGOs
3. Government Agencies
4. Research Institutes/Scientists
5. Administrators
6. Policymakers/people's representatives.

The research findings may be disseminated through the extension agencies. The research results will benefit both farming community as well as the promoting agencies, i.e. the Govt. or non-government organizations. Some of the researchable issues pertaining to some ITKs have been identified and presented in Table 2.

Table 2: Identification of Researchable Issues of Some Selected ITKs

Name of ITK	Purpose	Researchable Issues
Furrow opening in standing crops	Moisture conservation	Modification of implement with different types of serrated blades and also by introducing additional types
Nadi farming system	To collect run-off during *kharif* for life saving irrigation during drought spell or pre-sowing irrigation (*Palewa*) for *rabi* crops	Documentation and analysis of socio-economic aspect of present *nadi* system for its sustainability Evaluation of present *nadi* farming system
Mixed pulses as vegetative barrier	Resource conservation	Proportion of pulses as vegetative barrier Cost-effectiveness of the system
Stabilisation of gullies using sand bags	Gully control and run-off management	Soil conservation efficiency Strengthening of sand bags structure with different vegetative barriers
Application of white soil as lining material in farm pond	To work as a sealant material for lining dugout farm pond	Standardisation of application technique and economic feasibility for wider application Study on the seepage losses at different hydraulic heads
Wider row spacing in pearl millet	Moisture conservation and weed control	The plant population should be worked out for different situations
Rainwater harvesting in *kund/tanka*	The harvested water in *kund/ tanka* is used for drinking and establishment of tree	Research should be done on the use of stored water for arid horticulture Design of *tankas* for different geo-hydrologic conditions

(*Continued*)

Name of ITK	Purpose	Researchable Issues
Crop stubbles and residue management	Improve the organic matter and water holding capacity of soil	Effect on soil water conservation Yield advantage Alternate ways of composting and application
Brushwood waste weir	Safe disposal of excess run-off	Design and stabilisation of structure
Mulching in turmeric	To conserve moisture	Reduction in soil loss and improvement in soil quality and moisture availability Use of alternatives to Sal leaves as mulch material
Indigenous stone/ brushwood structure across the slope	To check soil loss	Shape and size of brushwood structure depending on the run-off and site conditions
Agave sp. as vegetative barrier	To reduce run-off velocity and to increase infiltration opportunity time	Different species of Agave can be evaluated Cost-benefit analysis.
Broad bed and furrow (BBF) practice	To harvest rainwater and dispose of excess water	Dimensions of the broad bed furrow need to be standardised Development of low cost tractor drawn implement for making BBF
Water harvesting and recycling	Rainwater harvesting	Recharging of water table Cost-effectiveness
Standardisation of recharging technique	Augmentation of groundwater	Design of filter and improvement in filtering efficiency with better filtering material. Effect of geology/soil formation on recharge
Set-row cultivation	For harvesting rainwater and maintaining soil structure	Quantification of harvesting additional water Improvement in soil health and crop yield
Summer/pre-monsoon tillage	Conservation tillage—to harvest early showers, facilitate timely seeding and weed control	Type of tillage tool Time and frequency of operation commensurate with soil type and depth
Ridge and furrow planting for modulation of overland flow	Conservation of rainwater, modulating excess water, control soil loss and boosting productivity	Fabrication and development of ridge former accommodating required row spacings and ridge cross-section
Pigeon pea field with *Gurr*	Reduction of run-off and soil moisture conservation	Impact of bullock and tractor made *Gurr* on crop performance Reduction in run-off in *Gurr* plot
Green manuring practice	To conserve soil moisture and improve nutrient status of soil	Proper time of sowing and turning into soil including feasibility of dry sowing Economic evaluation of green manuring cropping system

Name of ITK	Purpose	Researchable Issues
Application of tank silt	To increase the fertility and moisture holding capacity of soil	Mode and quantity of application of tank silt to different soils Cost-effectiveness of silt application (whether to apply to cereal crop and point application to horticulture crops) Quality of tank silt and its effect on crop

Conclusion

Many ITKs on *in situ* soil and moisture conservation are not adopted everywhere throughout India because of constraints in adoption and unawareness of the effectiveness of such practices. The present documentation process has definite bearing on the future course of action in framing new projects. This short-term documentation project may lead to the following future activities:

- Similar exercise can be undertaken to document the ITKs from all the agro-ecological regions of the country.
- The potential ITKs may be tested for their suitability and adoption in other agro-ecological regions as a dissemination strategy.
- The documented ITKs may be published/translated in all regional languages for the benefit of the farming communities.
- Validation of the ITKs is a logical step to qualify and quantify the effectiveness of these practices. Suitable modifications of the traditional practices through on-farm research would help in developing appropriate and acceptable technologies for different local environments.
- The effect of conservation measures on resource losses can be studied in detail through experimentation and use of stimulation model.
- As a policy matter the local ITKs should be in-built in the resource conservation programme.

REFERENCES

Agarwal, A. and Narain, S. (1999). Introduction, Dying Wisdom, Centre for Science and Environment, New Delhi, pp. 4-24.

Dividends from Soil and Water Conservation Practices (A Brief Review of Work done in Rainfed Eco-Regions). NATP, CRIDA, Hyderabad, p. 40.

Mishra, P.K. (2002). *Indigeneous Technical Knowledge on Soil and Water Conservation in Semi-arid India* (Eds: P.K. Mishra, G. Sastry, M. Osman, G.R. Maruthi Sankar and N. Babjee Rao). NATP, CRIDA, Hyderabd, p. 151.

Mishra, P.K., Sastry, G., Osman, M., Babjee Rao, N. and Maruthi Sankar, G.R., (2002).

Socio-Economic Constraints of Dry Farming in Tamil Nadu

7

K. Thangaraja, C. Karthikeyan and *J. Venkatapirabu*

Introduction

Dryland area receives less than 750 mm rainfall. Out of the net sown area of 136.18 m ha, the dryland accounts for 93.13 m ha (68.4%) and contributes 55 million tonnes of foodgrain production. This dryland agriculture from 93 m ha supports 40 per cent human population, 60 per cent of cattle heads and contributes 44 per cent of total foodgrain production in India (Kannaiyan et al., 2001). In Tamil Nadu, the area under dry farming constitutes 52 per cent of total cultivable area contributing to 40 per cent of total foodgrain production. Out of 5.5 m ha of net sown area, nearly 3.2 m ha are rainfed (Kannaiyan et al., 2001).

Crop selection for the dryland region, along with its management has a major impact on how efficiently the available natural resources are utilised to produce an economic yield. Though several factors are affecting crop production in drylands, rainfall is the important factor recognised, which is highly unpredictable in terms of onset of monsoon, recession and distribution. Unfavourable crop growth environment, limited choice of crops and varieties, poor soils, low cropping intensity, short growing season and low unstable productivity are the major concerns of the dryland farmers in Tamil Nadu (Selvaraju et al., 2001).

Majority of the dryland farmers are small farmers with scarce resources.

The poor resource base allows only a low input subsistence farming with low and unstable crop yield. The low productivity of agriculture in dry farming regions is due to the cumulative effect of many constraints for crop production. Such constraints need to be identified to overcome with suitable remedies. There is an urgent need to formulate strategies to increase the agricultural production in the dry farming tracts. The present study was conducted with the specific objective to identify the socio-economic and other factors influencing dry farming and suggest measures for its improvement.

Methodology

The present study was conducted in three revenue villages of Reddiarchatram block of Dindigul district in Tamil Nadu. The Reddiarchatram block was selected purposively as their main occupation was agriculture which depended on poor and erratic rainfall and a majority of the farmers of the block had been practicing dry farming. Three revenue villages having maximum area in dryland under cultivation from Reddiarchatram block were selected. A sample size of 90 farmers consisting of 45 maize growers and 45 sorghum growers was considered. The selection of 30 farmers was done at random in each village. Thus, 10 farmers were selected from each of the nine villages covering three revenue villages using simple random sampling technique. The data were collected with well-structured and pre-tested interview schedule. Percentage analysis was used for data analysis.

Findings and Discussion

Technological problems and suggestions perceived by dryland farmers

Perusal of the Table 1 reveals that the most important problems reported to have been faced by 78.88 per cent of the dryland farmers were non-availability of farm machineries in time (tractor drawn cultivator, maize sheller). Hence, it was suggested to improve the farm power in dryland area through government agency by more than half (53.33%) of the dryland farmers.

Non-availability of quality seeds and absence of location specific drought resistant/tolerant varieties in crop were felt as the next important problem by 72.22 per cent of the dryland farmers. For this, 46.67 per cent of the dryland farmers suggested that farmers should be supplied with adequate quality seeds through government, co-operatives or private companies in time and also release drought resistant variety through Agricultural University/Agricultural Research Station/Private Seed Companies, which could be a remedy to this problem.

Table 1: Problems and Suggestions Perceived by Farmers in Dry Farming (n=90)

Sl. No.	Perception of farmers	No.	%
A	**PROBLEMS**		
I	**Technological**		
1.	Non-availability of farm machineries in time (tractor drawn cultivator, maize sheller)	71	78.88
2.	Non-availability of quality seeds and minimum drought resistance/ tolerant varieties	65	72.22
3.	Minimum usage of fertilizers and plant protection measures	52	57.77
4.	Lack of knowledge, awareness about use of micronutrient and bio-fertilizers	51	56.66
II	**Socio-economic**		
1.	High wages demanded by agricultural labourers for weeding and harvesting stage	70	77.78
2.	Less frequency of contact with extension agency	64	71.11
3.	Unstabilized market price for their produce	60	66.66
4.	Low, uncertain and unstable farm income	52	57.78
5.	Lack of adequate credit facilities	44	48.89
III	**Climatic**		
1.	The onset and withdrawal of monsoons are highly variable and uncertain	62	68.89
2.	Lack of rainfall erratic and uneven distribution of rainfall	48	53.33
3.	Crop lodging due to heavy winds	32	35.55
IV	**Natural Resource Related**		
1.	Poor soil fertility status	48	53.33
2.	Weed is a major problem when the cultivable land kept into fallow land	34	37.77
3.	Soils have poor water holding capacity	32	35.55
4.	High temperature and low moisture reduced microbial activity	19	21.11
V	**Others**		
a.	**Extension**		
1.	Less number of field demonstrations and farmers meeting	52	57.77
2.	High yield gap between research and farmers field	47	52.22

(*Continued*)

Sl. No.	Perception of farmers	No.	%
b.	**Crop management**		
1.	Limited quantity of application of organic manure	43	47.78
2.	Uneven crop stand due to broadcasting of seeds	39	43.33
3.	Selection of traditional variety by farmers	32	35.56
B.	**SUGGESTIONS**		
I	**Technological**		
1.	Improve the farm power through government agency	48	53.33
2.	Supplying and releasing drought resistant varieties	42	46.67
3.	Increase the usage of fertilisers and plant protection chemicals	37	41.11
4.	Conduct meeting and demonstration in village level	33	36.67
II	**Socio-economic**		
1.	Improving labour saving technologies	52	57.78
2.	Regular visit by Assistant Agricultural Officers and other officials	43	47.78
3.	Fix reasonable market price for agricultural produce	40	44.45
4.	To develop Integrated Farming System (IFS) in dryland area	37	41.11
5.	Easy access to credit	34	37.78
III	**Climatic**		
1.	Crop selection and adjust the time of sowing	54	60.00
2.	Using various water harvesting techniques	39	43.33
3.	Raising wind break crop	25	27.78
IV	**Natural Resource Related**		
1.	Soil test based fertilizer application	41	45.56
2.	Promotion of dryland horticulture	30	33.33
3.	Summer ploughing	27	30.00
4.	Application of bio-fertilisers	15	16.67
V	**Others**		
a.	**Extension**		
1.	Increase the number of village meeting and field demonstration	43	47.78
2.	Improve the awareness and adoption of recommended dryland technologies	35	38.89
b.	**Crop management**		
1.	Farmers could use recommended level of organic manures	39	43.33
2.	To take appropriate sowing method	35	38.89
3.	To promote availability of improved variety	31	34.44

Minimum usage of fertilizers and plant protection measures was reported by 57.77 per cent of dryland farmers. It was suggested by 41.11 per cent of the dryland farmers to improve/ increase the usage of fertilizer and plant protection chemicals by dryland farmers, through these inputs supplied at subsidy rate by government co-operative society and agriculture depots.

Lack of knowledge, awareness about use of micronutrients and bio-fertilizers was felt as the next important problem by 56.60 per cent of the dryland farmers. So, as suggested by 36.67 per cent of the dryland farmers, village level extension officers should take efforts to conduct meeting and demonstrations, on the use of micronutrients and bio-fertilizers to crops.

Socio-economic problems and suggestions perceived by dryland farmers

An observation of Table 1 reveals that the first important problem reported by dryland farmers (77.78%) was higher wages demanded by agricultural labourers for weeding and harvesting operations. Dryland farmers were spending more money towards labour charges. About 57.78 per cent of the dryland farmers suggested to device and adopt labour saving tools and machineries to overcome this problem.

Less frequency of contact with extension agency was felt as a problem by 71.11 per cent of dryland farmers. Regular visit by Assistant Agricultural Officers and other officials may help to guide farmers in proper way as suggested by 47.78 per cent of the dryland farmers. The next important problem was unstabilised market price for their produce as reported by 66.66 per cent of the dryland farmers. About 45 per cent of the dryland farmers suggested that the government may decide the price for agriculture produce to get stabilised prices.

Low, uncertain and unstable farm income was felt as a problem by 57.78 per cent of dryland farmers. Development of dairy, poultry and agri-based allied activities to provide stable income in dryland area was suggested by about 41.11 per cent of the dryland farmers. The next important problem was lack of adequate credit facilities (54.44%). Most of the dryland farmers belonged to small and marginal farmers. They preferred to borrow credit from village co-operative society and other private banks/NGOs. For credit availability to the farmers, giving long-term loan by societies on the basis of their crop selection was suggested by 37.78 per cent of the dryland farmers.

Climatic problems and suggestions perceived by dryland farmers

An observation of Table 1 reveals that the dryland farmers reported three major climatic problems. The most important problem reported to have been faced by 68.89 per cent of the dryland farmers was the onset and withdrawal of monsoons were highly variable and uncertain. It was suggested by about 60 per cent of the dryland farmers to go for crop selection and adjusting the time of sowing based on the onset of monsoon.

The second important problem reported by dryland farmers (53.33%) was lack of rainfall. Watershed management in dryland areas may be the remedy for this problem as suggested by 43.33 per cent of the dryland farmers. Erratic and uneven distribution of rainfall are a general phenomena. By using various water harvesting techniques, rainwater can be effectively utilised for cropping in dryland area.

The third important problem was crop lodging in maize crop due to heavy winds as felt by 35.55 per cent of the dryland farmers. Raising wind break crop (Agathi) around the field bund to reduce wind velocity was suggested by 27.78 per cent of the dryland farmers.

Natural resource related problems and suggestions perceived by dryland farmers

An observation of Table 1 shows that the dryland farmers have reported few natural resource related problems. The most important problem was poor soil fertility status in dryland soil as reported by 53.33 per cent of the dryland farmers. So 45.56 per cent of dryland farmers suggested soil test based fertilizer application as a remedy for this problem.

Weed as a major problem when the cultivable land is kept fallow was felt by 37.77 per cent of dryland farmers. About 33.33 per cent of the dryland farmers suggested to promote dryland horticulture in dryland area and the farmers may be provided with seedlings at subsidised rate.

Dryland soils have poor water holding capacity was reported as the problem by 35.55 per cent of dryland farmers. Summer ploughing (May-June) improved the water holding capacity in dryland soil as suggested by 30 per cent of the dryland farmers. The next important problem was the prevalence of high temperature and low moisture with reduced microbial activity as reported by 21.11 per cent of dryland farmers. Summer ploughing, improved soil moisture storage and application of bio-fertilizers (Rhizobium, Azospirillum, Azotobacter) may be remedy for this problem as suggested by 16.67 per cent of the dryland farmers.

Other Problems and Suggestions Perceived by Dryland Farmers

Extension problems and suggestions

Dryland farmers reported two major extension problems and suggestions. The first important problem reported by dryland farmers (57.77%) was the conduct of less number of field demonstrations and farmers' meetings. Nearly half (47.78%) of the dryland farmers suggested for providing better extension service to know about various technologies. High yield gap between research and farmers field was reported by 43.33 per cent of the dryland farmers. Better awareness and adoption of recommended dryland technologies may reduce the yield gap as suggested by 38.89 per cent of the dryland farmers.

Crop management problems and suggestions perceived by dryland farmers

Dryland farmers had reported three major crop management problems and suggestions. Nearly half (47.78%) of the dryland farmers reported that limited quantity of application of organic manure was one of the problems. Increasing the number of milch animals, use of bullock to harvest required organic manure application may be a remedy for this problem as suggested by 43.33 per cent of the dryland farmers.

The second important problem reported by dryland farmers (43.33%) was uneven crop stand due to broad casting of seeds. About 38.89 per cent of dryland farmers suggested to avoid broadcasting of seeds particularly for sorghum and to adopt appropriate sowing method for this problem.

Selection of traditional variety by farmers was reported by 35.56 per cent of the dryland farmers as a constraint. Increasing the availability and access of improved variety among dryland farmers by Assistant Agricultural Officers and private company field officers may serve as a remedy for this problem was suggested by 34.44 per cent of the dryland farmers.

Conclusion

The findings of the study clearly reveal that the major technological problem and suggestion was non-availability of farm machineries in times and to improve the farm power status for common use in the village through government agency respectively. The major socio-economic problem given by farmers was high wages demanded by agricultural labourers for weeding and harvesting stage, so, improving the labour saving technologies, was suggested, climatic problem recorded was the onset and withdrawal of monsoon were highly variable and uncertain. So, selection of crop and adjust the time of sowing were suggested. Natural resource problem recorded was poor soil fertility status and the suggestion provided was soil test based fertilizer application, major extension problem recorded is less number of field demonstration and farmers' meeting. So, increased number of field demonstrations and meetings were suggested. The crop management problem recorded was limited quantity application of organic manure and hence use of recommended level of organic manure is suggested.

REFERENCES

Kannaiyan, S., T.M. Thiyagarajan, M. Subramanian, T.N. Balasubramanian and R. Selvaraju (2001). Dryland green revolution in Tamil Nadu: The Perspectives. Coimbatore: Tamil Nadu Agricultural University Press, pp. 10-12.

Selvaraju, R., T.N. Balasubramanian and T.M. Thiyagarajan (2001). Crop Management Strategies to Maximise the Productivity in Dryland Agriculture. In Dryland Green Revolution in Tamil Nadu: The Perspectives (eds.). Coimbatore: Tamil Nadu Agricultural University Press, pp. 93-95.

Emerging Trends in Gully Management—An Integrated Approach with People's Participation

8

Asim Biswas and *S.L. Budihal*

ABSTRACT

Over the years country's landmass has suffered from different types of degradation leading to the formation of wastelands. It is normally uncultivated, uninhabited bare lands with moisture stresses and poor quality of soil that is not conducive for regeneration of vegetation. Gullies, the advanced stages of rills, wide and deep channels indicate misuse, neglect and mismanagement of land over a long period. Rehabilitation includes control of extension of gullies, its reclamation, stabilisation and eco-restoration. A gully can be controlled by mechanical measures like contouring, terracing of sloppy lands or by making check dams, spillways, gully plugs, etc. Gully head cut is first to be controlled from further extension of its own. Afforestation with grassed waterways, bushes, vegetative barriers are some of the common vegetative measures to control gully. Reconstitution of the gullied ecosystem can be done by greening the sites through adoption of agro-technology like agro-forestry, pasturelands, silvipastural, hortipastural system, etc. Besides technological management, social management through group action performance and people's participation in rehabilitating gullied land is very important.

Hordes of gullies now remind us,
We would build our lands to stay;
And departing leaves behind us,
Fields that have not washed away.
Then when our boys assume the mortgage

On the land that's had our toil,
They'll not have to ask the question,
"Here's the farm, but where's the soil?"
—Anonymous
With apologies to Longfellow

Land, a non-renewable resource, is central to all primary production system. Over the years, country's landmass has suffered from different types of degradation. Biotic and abiotic pressures cause degradation of land. We cannot restore the soil once it is lost. We can hold and improve what is left, but topsoil scattered over the bottom of the ocean cannot be returned to the fields and pastures by any practical operation available to man. We cannot make water run uphill; but by slowing it's run-off downhill, we can provide a chance for percolation and store much of it in the reservoir of soil. Economic stability develops when soil is used intelligently and protected from erosion, excessive run-off and unnecessary wastage of rainfall. Wise land use means using land within its capability to keep it continuously productive. In order to do this, farmers need to control soil erosion. This is a public effort.

Unfortunately, soil, which is the physical basis of our agriculture, is one of the most unstable major resources. The topsoil, the principal feeding zone for plants, is very prone for erosion, washing away, by mainly three most active forces—water, wind and ice.

Nothing in Nature is absolutely static. Fortunately, eternal change is a fundamental law. Natural processes such as erosion have brought about the distribution and assortment of soil by wind, water and ice that have been immensely beneficial to all forms of life. Yet basic distinction should be made, however, between the rates of change brought about by nature, whereby soil building forces are substantially in equilibrium with those of destruction, precipitated by a man in the conquest of his environment. The one is inherently natural and the other must be considered artificial. What takes nature hundreds or even thousands of years to manufacture, man can and often does destroy almost overnight by haphazard improvident land use and husbandry.

Natural or normal geologic erosion is the erosion of land by water, wind, ice or other geologic agents in its natural state, undisturbed by human activity but accelerated erosion takes place when the eco-system is additionally disturbed, which is difficult to cope up by nature, by the action of humans and, sometimes, by animals. As the natural eco-system is disturbed, the balance between the soil formation and soil removal is greatly upset which marks the beginning of the hazards of erosion.

Among the agents of erosion, wind is directly responsible for sand and dust storms and is indirectly responsible for wave erosion along the shorelines of ocean and lakes. Water erosion is generally considered to manifest itself in two forms: (1) sheet washing and (2) gullying. It starts with detaching out of soil particles by splash action of raindrop and then gradually as sheet washing followed by rill, gully and ravine formation, which is the extensive network of gullies. All these forms represent the same process but different stages and generally, the next advanced stage will not appear until the preceding one has been underway for a considerable time. Gullies, however, occur sometimes without being preceded by sheet or rill erosion and conversely, sheet erosion is known to continue indefinitely without the formation of gullies.

The sheet erosion is the uniform removal of thin layers of soil from the surface of the earth caused by sheet flow or overland flow resulting in shallower soil with a progressive decline of crop yields. It is very difficult to judge until it will form rills by concentrating the surface flow along the slope. If it is kept unprotected, gradually it will concentrate over the same place producing wide and deep cracks forming gully like structure, which may assume bigger sizes at the advanced stages of rills. It can be easily managed by the normal tillage operation in the early stages. At the advanced stage of erosion when the depth of rills become more than 30 cm in depth are generally they called gully which cannot be obliterated by normal tillage operation.

On fields where erosion has been allowed to remain unchecked long enough for this to happen, not only the losses are accelerated but the expanse and difficulty of reclamation and control are greatly augmented.

When finger gullying or rills progressed to the point where the water channels are no longer interchangeable or easily obliterated, a state of gully erosion is said to have taken place. These gullies tend to widen and deepen with every rainfall. Quite frequently gullies in advanced stages may attain a depth of 50 feet or more, with proportionate width, in a remarkably short time and their encroachment on good land is not necessarily

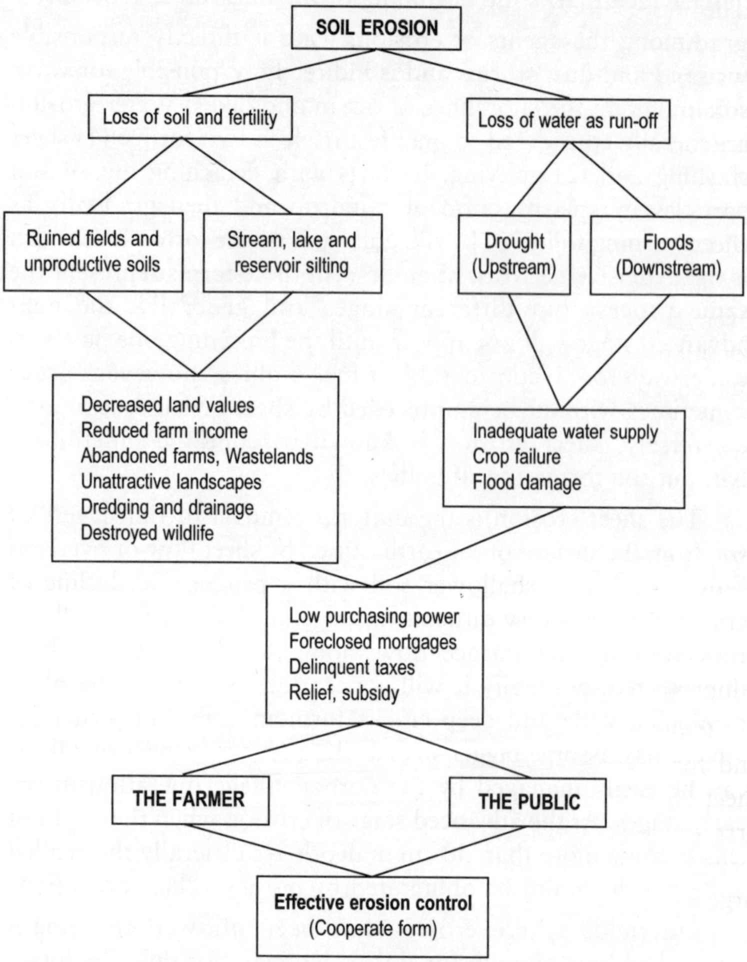

Fig. 1: Chart showing how soil erosion affects the general public as well as the farmer.

proceed by finger gullying. This is particularly true where the surface soil becomes loosened and overfalls develop in natural drainage.

There are generally two types of gullying. Ditch erosion occurs where head and sides of the gully are usually sloping and erosion occurs at the head, sides and the bottom of the gully in varying degrees by the action of water, freezing and thawing. The depth and shape of ditch or channel gully depends on gradient, rainfall, size and condition of the contributing watershed but chiefly on the soil. If the underlying material is soft and easily cut into deep, straight-walled and U-shaped gullies are formed. Where the subsoil consists of relatively un-absorptive clay, shallow V-shaped gullies usually result. Where hard rock occurs near the surface, the incisions will be shallow but usually wide.

Whereas the waterfall erosion is caused by water falling over the edge o f a gully or ditch bank and is often responsible for many of the deepest gullies, the falling water undermines the edge of the bank, which caves in and waterfall moves upstream. The undermining action of the waterfall goes on rapidly if the subsoil is of easily erodible type. Gullies formed by waterfall erosion may extend back through almost level land and their growth usually depends on the fall, type of soil and size of drainage area furnishing water, rather than on the slope of the land. Both types of gully erosion are often found in the same gully. Lateral gully tends to form from the main gullies and ultimately a network of gullies develops in advanced stages, which is also known as ravines.

Gully erosion is much more noticeable than sheet erosion and for this reason is generally considered more serious than sheet erosion. This, however, is a false impression because the first and most serious damage results from sheet erosion. The formation of gully erosion can be retarded and controlled to a large extent by preventing excessive sheet erosion.

Usually a good way to control a gully is to plant the entire ravine to stabilising vegetation. However, even where it is impractical to plant the entire gully, the head must be stabilised at last. The type of vegetation and the procedure necessary for its satisfactory establishment will vary considerably according to soil, climate and topographic conditions.

Run-off from a gullied field, however, must be handled carefully, particularly during the period when vegetation is being stabilised in or around the gullies. Run-off hazards depend on the size and condition of the watershed. A heavy concentration of water damages or washes out the plantings. Temporary dams, porous brush or wire barrages and diversion channels will help in many places in the establishment and satisfactory growth of plantings. If volume of run-off is too great for effective vegetation control, permanent structure must be utilised, including especially the use of diversions to prevent upslope run-off from entering the head.

From the standpoint of economy and practicability, the principle methods of handling run-off for gully control, in the order of importance, are the retention of as much of the rainfall on the watershed as may be possible; the diversion of run-off away from the gully, particularly from the head and the safe conveyance of run-off through the gully. The involvement of the large number of resource poor, technology-ignorant farmers need a cooperative move for water and ultimately the gully management.

Frequently, the practical solution will be a combination or co-integration of these methods with people's participation. Whatever methods are used should be planned and executed carefully. Haphazard and unplanned work leads to failure or even an aggravation of gully conditions. Until vegetation is well established it is better to exclude livestock grazing.

Generally, run-off or rainfall can be reduced materially through the use of ordinary soil and water conservation measures. Effective prevention and control of soil erosion and adequate conservation of a rainfall in a field, on a farm or ranch, over a watershed, or on any other unit or parcel of land require the use and treatment of kinds of lands comprising that area in accordance with the individual needs and capabilities of each different area of any important extent. The effective conservation practices have not yet been developed for all the problems and combinations of problems relating to the contrasting agricultural lands. It is necessary, therefore, to seek constantly for cheaper and more efficient ways of conserving soil and water. These will continue to be sought since man has not yet acquired all the facts relating to the land. Vegetative and mechanical conservation measures have been efficiently joined and applied as part of a complete conservation treatment for the land, the results on intertilled cropland probably have averaged about 85 to 90 per cent effective. Sometimes, the results are nearly or even 100 per cent effective, as under an excellent cover of the grass or forest.

A number of mechanical and agronomic measures are generally followed for the conservation of soil and moisture. Making the bunds across the slope along contour line or along the predetermined longitudinal gradient known as graded bunds using of local material are very common practices for soil and moisture conservation. Sometimes the graded boarder strips are made in between the contour bunds or graded bunds. The construction of series of platforms or step-like structures by slope cutting and levelling along the contour lines, known as terracing, are very common method in the regions of high sloppy land condition. Hill type bench terrace, irrigated bench terrace or orchard bench terrace are very common depending on the purpose or based on the slope it is of different type like bench terrace with level top or inward or outward sloping or Puertorican type of bench terrace. Sometimes making compartments in the field by compartmental bunding is also a common soil and water conservation practice.

Adaptation of contour farming is a best cultivation practice for conserving soil and water. Growing of crops in a close spacing or mixed cropping is very effective in conserving soil and water. The different strip cropping like contour strip, field strip, wind strip or permanent or temporary buffer strip cropping help in controlling run-off and conserving the soil and water. Making of vegetative barriers by different grasses or shrubs control the water movement conserving soil and water too. Different planting techniques like broad bed and furrow system helps in conservation of moisture in *in situ* conditions. Complete coverage of land with vegetation whether grasses, legumes, shrubs or trees is an ideal condition for absorption of rainfall and reducing rain splash erosion. Vegetation helps in reducing the direct impact of raindrops on soil surface and also reduces the speed of run-off greatly, allowing more time for penetration into earth reducing the loss of soil. The rotation of crops with grasses and legumes effectively controls erosion during the period of their growth on a given area and lessen erosion for some time thereafter. All these measures ultimately help in the control of soil erosion and development of different erosion structures.

In control of gullies, run-off usually should be diverted from the head of the ravine before control measures are installed within the channel. By terracing the drainage area above the gully head, much of the run-off can be diverted to other points for safe disposal.

Small and medium gullies with small to medium or uncultivated watersheds frequently can be stabilised by placing a series of earth fills, spaced according to the steepness of the land, across the gully channels, with grass or other holding vegetation established over the fills as quickly as possible. On the tableland at the upper part of the watershed area and the terraced land, wide gully hump-tops are transformed to good agricultural field. Different land use options and agronomical measures are to be adopted for resource utilisation on sustainable basis. The agronomical practice such as contour cultivation, i.e. all operations of tillage and planting are carried out along contours for moisture conservation.

However, under excessive heavy storms severe erosion damage may occur. Erosive velocities are reduced by constructing a series of checks which, in time, transform the longitudinal gradient of the gully from a uniform, steep slope to a succession of steps with low risers and long flat. These checks are usually of temporary character and serve to hold the fill and prevent washing while vegetation for permanent control is becoming established. For filling up the gully completely, additional checks are later build midway between the original structures. The conventional method for protecting the head bank heretofore recommended has to be cut back under the bank and pack the cavity with straw and brush to protect further development.

A surer way of protecting over-falls is to cut back the head bank to slope of about 30° and install a flume of ordinary planks, creosoted planks, or galvanised iron. These grades should be obtained by excavation, if necessary, or assured by the construction of checks. In lieu of flume, sometimes the head bank may be cut back as before, the sides smoothened and the channel lined with tough sod, rubble masonry or concrete reinforced with chicken wire mesh.

The best way to cure head over-falls, next to diverting the flow, is to build a permanent, watertight structure just below the head bank and count on this to hold the soil that is washed in from above it.

The head banks of gullies in field on which terraces are built lose their power to wreak further damage because the run-off is permanently diverted to other outlets. This is the most satisfactory way of controlling head erosion, since the gully may be simply filled in and forgotten. If the circumstances are such that terraces are impracticable or are not otherwise needed, a diversion ditch above the heads of a series of gullies is frequently used to concentrate the flow in a single channel where it may be more easily dealt with.

Growing of two or more crops together in a systematic manner helps in achieving higher productivity per unit time and area and the soil conservation come as a bonus. An augmentation of productivity on sustainable basis is possible with choice of crops and cropping pattern, appropriate agronomic practices and judicious use of limited available water.

Grassland development and afforestation for rehabilitation of gullied lands indicates versatility of this land use since vegetation is intrinsically benevolent to land and environment. It reconstructs the entire eco-system through site amelioration, greening of sites, return of wild-life and enrichment of high floristic and faunal diversity, control of water pollution in addition to fuel, fodder and fibre production. The essential factors are good nursery practice, suitable planting techniques, ground preparation, choice of suitable species and planting operation.

Similarly, grassland development, silvi-pasture development are practices for land in the upper region of gullied land. Utility of a tree and the grass combination in the silvi-pasture system can be adequately enhanced by judicious selection of plant species, compatible plants species adapted to the edaphic, climatic, topographic and biotic condition meeting local needs. At different gullied land area the combination of a different species of trees and grasses produces very good results in biomass production without causing further land degradation. At the same time a very good return may be expected from the goats reared by grazing them in that land area.

A grassland in good condition consists of many different kinds of plants, of different degrees of palatability, in different patterns of seasonal development and many varieties of growth form. A vegetation of good condition would contain predominant plant species of perennial character. On the other hand, predominance of vegetation is indicative of degraded state. The biotic interference, controlled succession takes two to five years for occurrence of perennial palatable grasses and 10-15 years mesophytic trees to replace grasses. These trees are the highest form of vegetation and economic species found in tropical climate. The density of meditation and biodiversity can be improved by adoption of moisture or water conservation practices.

Management Perspective

The degraded land management is social process of planning, organisation and actuating through a co-operative group action for securing maximum benefits from the natural resources, viz. land, water, meditation, animal including human with minimum effort so as to secure maximum prosperity and happiness of stakeholders, user groups directly and other people and government indirectly. The management of such type land requires generation of cost-effective practices and their application through participatory approach.

Depending on the inhabitants of the area social homogeneous group, viz. users group, self-help groups and gender development and welfare groups are to be constituted by way of capacity building, i.e. motivation, actuation and control.

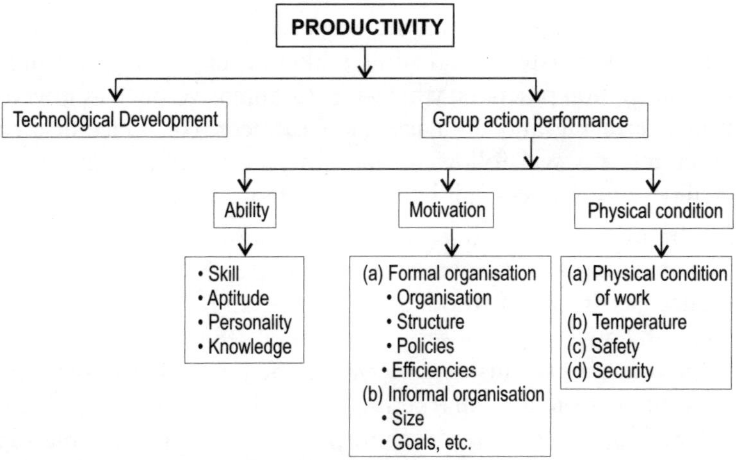

Fig. 2: Factors of productivity in degraded land of gullied watershed as per principle of management.

Thus, application of technologies of research for the natural resource of land and water when applied by the new modus operandi will produce more prosperity and happiness in the gullied lands on sustainable basis.

The productivity is constituted by technological development and group action performance, which are again function or ability, motivation and physical condition. Snag in either of the factors will hinder productivity. Earlier efforts were mainly on technical innovations, which resulted in techniques and practices for natural resource conservation and utilisation. When these were transferred for implementation, their success did not come up to mark. That means there occurred failure on part up implementation and maintenance for sustainability. The new paradigm lies in changing modus operandi from top to bottom to new style that is from bottom-up approach. The modus operandi required capacity building and managing and not administering the things.

In gullied waste land management, both technological and social management parts are necessary and complementary. The division of work is drawn that clearly specifies who has to do what. The government's role is in creating conditions for fulfilling the physical needs of the people. This situation will help induce participant's eagerness in the resource poor and technology ignorant subsistence farming stakeholders. The landform transformation practices, which are cost intensive, such as inward bench terracing and drainage line treatment, etc. Once created other reforms will follow up fast. In participation mode other facilities and services can be easily created by group actions, etc.

In practice, usual expression is shortage of funds for implementing watershed treatment measures. While technologies reduced cost and method of implementation, there lies great scope in convergence of services of other department, viz. road, irrigation, animal husbandry, welfare activity, health, security, communication and marketing, etc. This needs a good co-ordination, among interdepartments and social transformation and for cooperative movement.

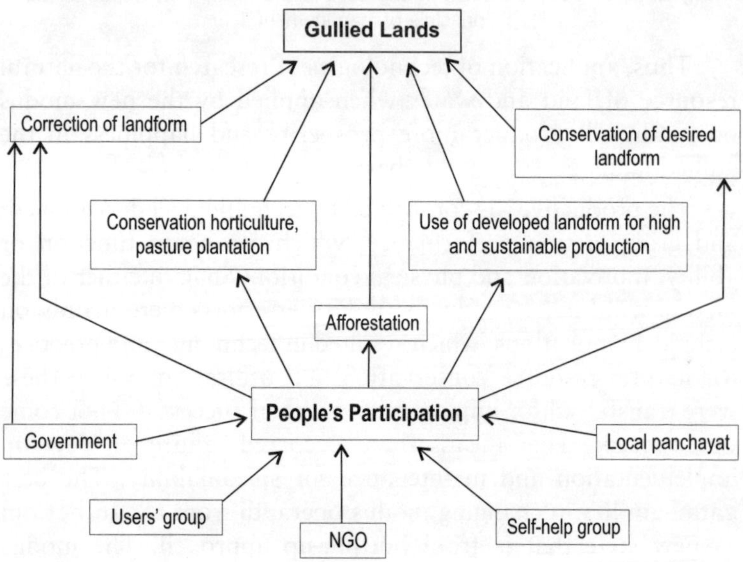

Fig. 3: Rehabilitative work/actions for management of gullied lands.

Depending on the inhabitants in the watershed social homogenous groups, viz. user groups, self-help groups and gendered dependent and welfare groups are to be constituted by way up capacity building, i.e. motivation, actuation and control. Thus, application of technologies of research for the natural resource of land and water when applied by the new modus operandi will produce more prosperity and happiness in the gully degraded land on sustainable basis.

REFERENCES

Bhushan, L.S. and Yadav, R.C. (2000). Management of Wastelands with Special References to Ravines. *Indian Farming*, February 2004, pp. 22-31.

Hugh Hammond Bennet (1955). *Elements of Soil Conservation*, McGraw-Hill Book Company Inc., NY, USA.

Prajapati, M.C., Nambiar, K.T.N., Puri, D.N., Singh, J.P. and Malhotra, B.M. (1993). Fuel and Fodder Production in Yamuna Ravines at Agra. *Indian J. Soil Conservation*, **21**(3), pp. 8-13.

Quincy Claude Ayres, C.E. (1936). *Soil Erosion and its Control*, McGraw-Hill Book Company Inc., NY, USA.

Impact Evaluation of DPAP Watersheds in Tamil Nadu

9

P. Balaji and *C. Velavan*

Introduction

Rainfed agriculture in India is characterised by low productivity, degraded natural resources and widespread poverty. Most of the millions of people living in our country depend on agriculture and natural resource management for their livelihoods, so development planners are eager to implement productive, environmentally sustainable land and water management. It is in this context, the concept of watershed development has been introduced in our country. In order to bring about a balanced and sustained development of Drought Prone Areas, Government of India, instituted programmes like DPAP, DDP and IWDP. All these programmes are based on the watershed basis.

Salem district is one among the 16 districts selected for the implementation of Drought Prone Area Programmes in Tamil Nadu. These districts have been declared as drought prone districts by the Government of India on the basis of their characteristics like low extent of irrigated area, low rainfall, erratic distribution of rainfall and high periodicity of drought.

What is watershed?

A watershed is a geographical area that drains to a common point, which makes it an attractive unit for technical efforts to conserve soil and maximise the utilisation of surface and subsurface water for crop production. It is an ecosystem or bio-geographical unit in which the interdependence is internalised. Watershed development projects are designed to harmonise the use of water, soil, forest and pasture resources in a way that conserves these resources while raising agricultural productivity, both through *in situ* moisture conservation and increased irrigation through water harvesting.

Watershed development has been conceived basically as a strategy for protecting the livelihoods of the people inhabiting the fragile eco-systems experiencing soil erosion and moisture stress. The aim has been to ensure the availability of drinking water, fuelwood and fodder and raise income and employment for farmers and landless labourers through improvement in agricultural production and productivity. In Salem district under this programme 15 watershed areas, as given in Table 1, have been treated at the allocated cost of Rs. 300 lakh of which Rs. 270 lakh have been received at 50:50 ratio from both Central and State Government since 1995-96. The details of the watersheds covered in Salem district are given in Table 1.

Table 1: Details of DPAP Watersheds in Salem District

Name of the block/watershed	Total area (Ha)	Total area covered (Ha)
Kadayampatti		
Kanavaipudur	855	500
Govindapuram	840	500
Bommiampatti	735	500
Mecheri		
Kovilur	980	500
Banapuram	505	505
Thethigiripatti	510	510
Nangavalli		
Chinnasoragai	590	500
Periasoragai	570	500
Gonur	510	510
Konganapuram		
Thangayur	540	500
Ermapatti	520	520
Koranampatti	530	500
McDonald Choultry		
Ernapuram	605	505
Egapuram	610	510
Kannanderi	515	515

The overall project performance is given in Table 2.

Table 2: Performance of the DPAP in Salem District

Particulars	Units	Physical		Financial (Rs. Lakhs)	
		Target	Achievement	Target	Achievement
Homestead planting	Ha.	420	420	3.46	3.40
Horticulture extension	Ha.	1714	1714	46.78	41.03
Calf rearing activities	Nos.	616	616	20.68	18.18
Community farm development	Ha.	15	15	2.40	1.08
Tankshore plantation	Ha.	25	25	0.93	0.91
Drip demonstration	Nos.	46	46	4.95	4.16
Major check dams	Nos.	90	90	40.50	37.57
Community borewells	Nos.	18	18	9.77	8.30
Percolation pond	Nos.	7	7	14.80	11.48
Tank strengthening and desilting	Nos.	13	13	17.90	16.57
Land levelling	Ha.	54	54	5.54	5.50
Improvement of inlet channel	RM	2165	2165	17.14	15.03
Community threshing floor	Nos.	51	51	28.18	21.32
Loose rock check dams	Nos.	30	30	1.20	1.20
Contour check walls	Nos.	292	292	16.17	16.17
Minor check dams	Nos.	40	40	3.60	3.60

What should be evaluated?

In evaluating watershed projects, it is more important than the aspects that are to be evaluated.

Physical development of the watershed: The most immediate objective of watershed development is to arrest soil erosion, improve moisture conservation and harvest rainwater. These may be evaluated. If the investment in soil and water conservation is to be sustainable, linkages have to be developed so that the improved resources and the dynamic local institutions can carry the development process forward.

Capacity of the stakeholders: Members of the watershed community, the agency implementing the watershed development projects and the government (or other donor) agencies dealing with watershed development issues, all have to have their capacities built. In the case of capacity building evaluating the number of trainings imparted is insufficient; the quality of training and skill development will determine whether capacity is actually built. Here proxy indicators will have to be employed that measure changes in attitudes, in skills, in information and awareness and inabilities, improved capacities should get reflected in how things are done and these latter may be measured.

Creation of sustainable institutions to manage the watershed resources: The desirable features of the institutions being promoted should be listed and these should be evaluated. There are objectively verifiable indicators that can measure such characteristics. For example, it may be stated to be desirable that the institution should be sensitive to equity. This can be measured from the decisions taken in institution, recorded in the minutes of meetings.

Improvement in the socio-economic status and quality of life: Watershed development leads to increased incomes of the members of the community, particularly of the poor men and women. This economic and equity aspect of the programme should be evaluated.

Monitoring and Review Plan: Every project or programme should have a monitoring and evaluation plan. It would state what results are to be achieved at different stages of the project/ programme, what the indicators of success are, how and when they are to be measured and who is responsible for measuring them.

Objectives

(i) to study the impact of water resources development in Salem district;

(ii) to study the impact on agricultural productivity; and

(iii) to study the Overall Impact of the Watershed Development Activities.

Result and Discussion

The impact of various watershed development activities on water resources development was analysed (Table 3). It could be seen from the table that in all the watershed treated areas, the water coloumn of wells is higher than the untreated wells. It is evidenced that the irrigation intensity is higher than the untreated villages. This lucidly shows watershed development activities help increase the water resource potentials of the region.

On the basis of the data collected from the sample wells, it was found that the water level in these dug wells has risen in the range of 0.5 m to 1.5 m. This is mainly due to the construction of percolation ponds and major and minor check dams (Table 4).

Duration of water availability over the year in the wells indicates that due to watershed intervention the perenniality of wells is higher in the case of watershed treated villages than the untreated villages (Table 5).

The impact of watershed intervention on cropping pattern, crop productivity were analysed. The data on cropping pattern indicate that the cropping intensity is relatively higher in the case of watershed treated areas. For example, in Kanuvaipudur watershed, the cropping intensity is worked out to 185 per cent and 105 per cent respectively for the watershed treated farms and untreated farms. The Crop Diversification Index (CDI) was worked out based on the area under each crop in different samples for both watershed treated and control farmers. The CDI values nearer to 1 indicates that there is complete diversification. It is evidenced that the CDI is higher in the case of watershed treated farms than the control confirming watershed treatment activities help farmers to diversify their farms activities (Table 6).

Increase in crop yield has been reported by farmers though the extent of increase varies from crop to crop and watershed to watershed. Index of crop productivity was calculated by using farmers' yield data and normal yield of a crop as per the standard package of practices to evaluate changes in crop productivity. Percent increase in crop productivity level varies from 2.5 per cent for groundnut in Ernapuram watershed to 27 per cent for groundnut in Chinnasoragai watershed (Table 7).

In both the watershed treated areas and control villages, households need cash income to supplement agricultural production (crop production, livestock and trees) to purchase market purchased commodities, to pay for social obligations, school fees and health care, etc. Most of this income is generated through off-farm and non-farm income activities. Table 8 provides worked out cash income and relative importance of different sources of income. It is evidenced that the income from crop production is higher in the case of watershed treated farms than the control. In addition to crop and livestock production, off-farm and non-farm activities, tree growing enable the households to derive additional income through the sale of wood, timber, seeds and fruits, etc. Households participate in off-farm and non-farm income activities particularly during slack periods to get additional income for their subsistence needs. The share of off-farm and non-farm income sources is higher in control villages than the watershed treated areas. In order to meet cash expenses, the households participate in off-farm wage labour and non-farm income activities such as employment in rice and oil mills, etc. (Table 8).

Of the different factors studied, the watershed development activities exert more impact on groundwater recharge followed by soil and water conservation and improvement in soil fertility in almost all the watersheds (Table 9). It is evidenced that the watershed development activities help to increase cropping intensity, alter cropping pattern as evidenced by higher mean scores.

Table 3: Impact on Groundwater and Irrigation Potential

Particulars	Water column of wells (m)	Net area irrigated (Ha)	Gross area irrigated (Ha)	Irrigation intensity (%)
Chinnasoragai				
Treated	2.58	2.13	2.54	119.24
Control	1.72	1.69	1.69	100.00
Ernapuram				
Treated	2.35	1.44	1.68	116.67
Control	1.95	1.22	1.22	100.00
Kanuvaipudur				
Treated	2.66	2.31	2.86	123.87
Control	2.18	2.02	1.96	97.03
Kovilur				
Treated	2.48	2.92	3.75	128.42
Control	2.20	2.23	2.42	108.52
Thangayur				
Treated	2.50	2.20	2.80	127.27
Control	2.10	1.98	2.00	101.01

Table 4: Rise in Water Level due to Watershed Development Activities

Particulars	Rise in water level (m)	Average water level rise (m)
Chinnasoragai	1.0-1.5	1.25
Ernapuram	0.5-0.75	0.60
Kanuvaipudur	1.0-1.50	1.30
Kovilur	0.50-1.0	0.70
Thangayur	0.5-0.75	0.60

Table 5: Duration of Water Availability in the Wells Influenced by Percolation Pond and Check Dams (Number of wells)

Particulars	12 months	9 months	6 months
Chinnasoragai			
Treated	5	3	2
Control	4	4	2
Ernapuram			
Treated	4	4	2
Control	2	3	5
Kanuvaipudur			
Treated	4	3	3
Control	3	4	3
Kovilur			
Treated	5	3	2
Control	1	4	5
Thangayur			
Treated	3	4	3
Control	3	3	4

Table 6: Impact on Cropping Pattern

Particulars	Cropping intensity (%)	Crop Diversification Index (CDI)
Chinnasoragai		
Treated	180.00	0.83
Control	98.78	0.70
Ernapuram		
Treated	109.37	0.74
Control	100.00	0.60
Kanuvaipudur		
Treated	185.00	0.96
Control	105.00	0.84
Kovilur		
Treated	110.00	0.81
Control	100.00	0.70
Thangayur		
Treated	124.00	0.64
Control	86.64	0.50

Table 7: Index of Crop Productivity for Major Crops

Particulars	Groundnut	Rice	Sugarcane	Cotton
Chinnasoragai				
Treated	0.71	0.72	0.98	0.77
Control	0.56	0.64	0.94	0.65
Per cent difference	27	13	4	18
Ernapuram				
Treated	0.82	0.78	0.77	0.81
Control	0.80	0.69	0.69	0.78
Per cent difference	2.5	13.04	11.59	3.86
Kanuvaipudur				
Treated	0.70	0.72	0.85	0.72
Control	0.65	0.60	0.78	0.63
Per cent difference	7.69	20	8.9	14
Kovilur				
Treated	0.81	0.76	0.77	0.69
Control	0.78	0.66	0.75	0.58
Per cent difference	3.8	15.1	2.6	18.9
Thangayur				
Treated	0.77	0.83	0.65	0.66
Control	0.65	0.73	0.63	0.58
Per cent difference	18	14	3	14

Table 8: Impact on Income of the Farm Households

Particulars	Crop income (Rs.)	Off-farm and non-farm income (Rs.)	Livestock income (Rs.)	Household income (Rs.)
Chinnasoragai				
Treated	25198.03	2976.00	7289.00	35463.03
Control	14104.91	2166.00	4846.64	21117.50
Ernapuram				
Treated	14896.00	3215.00	6546.00	24657.00
Control	14106.74	4164.00	2118.64	20389.38
Kanuvaipudur				
Treated	25211.76	4345.47	6968.61	36525.84
Control	14412.21	4199.00	2088.33	20699.54
Kovilur				
Treated	14840.00	5164.00	5498.00	25502.00
Control	14150.00	5965.00	3124.00	23239.00
Thangayur				
Treated	16098.00	4124.00	6128.00	26350.00
Control	14956.00	4010.00	2088.00	21054.00

Table 9: Overall Impact of the Watershed Development Activities

Particulars	Chinna-soragai	Ernapuram	Kanavai-pudur	Kovilur	Than-gayur
Groundwater recharge	45.9	54.6	53.6	52.18	45.37
Yield increase	0.9	1.2	0.7	0.7	0.8
Increase in cropping intensity	9.8	14.5	12.7	12.0	10.8
Resource saving	12.8	15.9	19.8	15.9	15.2
Soil and water conservation improved	30.7	43.1	40.9	36.1	38.9
Soil fertility improvement	36.3	37.9	36.8	42.1	32.8
Farm diversification	1.4	2.1	1.5	2.5	3.7
Helps to alter cropping pattern	5.4	3.4	3.2	4.7	3.0
Increased fodder availability	0.5	0.8	0.6	0.4	0.8
Increased fuelwood availability	0.7	0.7	0.5	0.8	0.9

Note: Figures indicate worked out mean score employing Garett ranking technique.

Farm Watershed Management Strategies for Arid Agro-Ecosystem

10

G.D. Singh

For the sake of sustainable growth in agriculture and to achieve food security, we should improve the productivity of the rainfed farming areas, which cover nearly 63 per cent of the total cultivated area. We should take advantage of the diversity of growing conditions prevalent in the country, which enable the cultivation of a wide range of crops, vegetables, fruits and flowers. Horticulture and related farming systems provide opportunities for improving nutrition security at home and export earnings abroad. Watershed management technology may incorporate measures to harvest and conserve rainwater, so that it can be used for sustainable production of agriculture and arid fruits. Maximum emphasis is to be placed on rain water integrated management of water resources and use of techniques of micro irrigation.

Experiencing a setback to natural resources in irrigated areas, during green revolution now the time has come to improve productivity of arid agro-eco-system which cover a large percent of total cultivated areas. These are the areas which are still free from vagaries of green revolution, yet offer superb advantage of the diversity of growing conditions. Obviously arid fruit culture may provide a lot of opportunities for improving nutritional security and export potential.

Major dryland areas apart from abiotic problems are doing well on national agricultural front. Apart from abiotic problems of adverse climate, lack of water available for crop production is the main challenge towards sustainable production of crops or fruits in dryland watershed areas. As such judicious management of watershed may be an appropriate solution for sustainable agri-horticulture.

Practically rainwater harvested in *khadin* or Farm ponds prove valuable asset to be used as life saving irrigation in agriculture or horticulture. Certain cultural practices and conservation technologies have been found effective for integrated watershed management for sustainable agriculture on watershed basis, such as rainwater treatment, contour farming, *in situ* conservation, management of abiotic stress, farming system management, planting technique, cultural operations and efficient micro irrigation.

Rainwater Treatment

Rainwater either stored in root zone of soil or collected on surface is successfully integrated for diversified agriculture and fruit crop production in dryland areas having poor quality groundwater by reducing irrigation needs to one or two irrigation. Also used as life saving irrigation from harvested rainwater. Rainwater harvested in tanks has to be used for supportive irrigation in fruits and agri-horticulture. Root zone of fruit crops may be saturated by moisture with practices like bio-mulching, strip cropping, compartment bunding and ridge and furrow making on land surface. *In situ* conservation by such techniques generally increases water available for crops. In a NATP project strip cropping study in Rajasthan maximum returns were recorded under pearl millet and cluster bean grown in strips in 3:1 ratio with maximum moisture (Singh, 2002).

Compartmental bunding is responsible for efficient use of rainwater and increasing quality and productivity of crops thereby integrate water use in watershed areas.

In dryland areas rainwater is harvested on the farmland in the natural depression as farm pond or *khadin* and tanks properly designed with cemented catchment areas with dimensions befitting annual rainfall obtaining in the area. An underground water-harvesting tank of 25000 L capacity with a catchment area having 36 feet diameter has been designed by Government of Rajasthan DWSC Department for annual rain-fall of 300 mm. In Balesaria watershed in Rajasthan formation of CVB, contour bunding. *Nadi* and water course treatment by watershed users committee helped harvesting of rain leading to rise in the water table by 5-15 feet during 7 years. Rainwater conservation methods in the watershed are much efficient ways of integrating water management for superb quality fruits. Land surface, roof top, *in situ*, flood water conservation and tanks are promising methods of rainwater management.

Contour Farming

Contour farming is a system of farming interventions such as cultivation, bunding and planting on the contour lines on moderately sloppy terrain in watershed areas. Contour bunding is a promising management system in arid eco-system in the sense that establishment of contour vegetative barriers are difficult on account of drought conditions. However, in the areas having good rains, Contour Vegetation Barrier of *Sachhrum munja* or *Cenchrus ciliaris* are proving promising method of soil and water conservation. In semi-arid areas, a weed *Sorghum helepense* has been found successful for construction of CVB. Provision of fodder for domestic animals from well established grass strip could be an additional benefit of contour vegetative barrier. Contour bunding and CVB of *S. helepense* are reported to increase production of *mateera* with *kharif* crops at Fatehpur-Shekhawati. For more effective moisture utilisation planting may be done near the CVB line just on uphill side of the topography. Also all efforts should be done to plant fruit plants on the contour lines to harvest maximum benefit of soil and water conservation in case CVB are not established. In a watershed system, well established contour bunds and CVB generally help fruit plants in three ways:

(1) Increase moisture availability for the individual fruit plants.

(2) Control soil erosion, thereby improve soil fertility.

(3) Increase productivity of fruits and agri-horti system.

In a NATP project study the highest soil moisture and yield of green gram was recorded under contour bunding followed by CVB of *S. helepense* as compared to *Cenchrus ciliaris* (Singh, 2002).

In Situ Conservation

In the arid eco-system of Thar desert small *mateera* (called loya) are cultivated on barren sand dunes with the help of efficient water management technique developed by local people as ITK, called *badi* cultivation. Generally 2 × 2 m compartment bunds are prepared and in each compartment two pits (20 × 20 × 20 cm) are made for putting 5 seeds of *mateera* with some water (2 lit.) for germination in March-April. The seeds germinate but plants grow very slowly but as such are protected in summers by a cover of thorny plant residue. With the onset of monsoon, growth of *mateera* is very fast and fruiting begins in July. The immature small fruits of *mateera* are harvested for early vegetable to fetch high price. These loya are very famous for delicious vegetable in off-season in dry areas.

Rootzone of the fruit crop may be saturated by water following *in situ* conservation practices like conservation tillage, bio-mulching and compartment bunding. Rainwater is harvested on the farm land in a natural depression as *khadin* or farm pond and sub-surface tank. Use of bio-mulching by weed or straw gave promising response as far as available soil moisture in *kharif* and *rabi* is concerned. However, the effect of bio-mulching was found more prominent in winter season as compared to *kharif* season. In a study in *in situ* moisture; use of *Tephrosia* sp. bio-mulch recorded significantly more grain yield of pearl millet and mustard as well as soil moisture over control (Annon., 1995). A promising moisture saving technology of watersheds say; contour trenching (*kunda*) recorded maximum incremental crown diameter (ICD) in *Prosopis cineria* and moisture conservation over other treatments. Water use efficiency by trenching technology is further accentuated by use of grassy straw mulch around the trees (Singh, 2002).

In arid and semi-arid areas intense radiation and excessive evapo-transpiration causing drastic losses of moisture, certain efficient mulches like plastic film, vegetation (green), straw and grasses may be used to spread on the soil surface near the trees or between the rows of fruit trees. Mulches help in moisture conservation, stabilising soil temperature, checking evapo-transpiration, add soil organic matter and useful micro climate.

Management of Abiotic Stress

Crop productivity and its quality is adversely affected by abiotic vagaries like hot winds, frost and other weather factors. Prevalence of hot or cold winds at maturity results to shrivelled seeds, fruits cracking and poor quality products. Crops may be frequently spared from abiotic stress by judicious use of wind control strips, shelter belt, mulches, frost control and laying of contour vegetative barriers for stress management. Practice of *kana bandh* in dry areas affected by wind erosion has been found promising in sustaining soil moisture and soil carbon (Table 1). Plastic mulch and plant growth hormones also reported to act as a shield against adverse abiotic effects on crop productivity.

Table 1: Effect of Wind Erosion Control *kana bandh* on Moisture Conservation and OC in a Watershed in the Month of September (Singh, 2002) at Fatehpur Shekhawati

Kana bandh	Soil moisture in Sept.	Organic carbon in Sept.
Ridge/furrow	1.28	0.33
Khejri sticks	2.27	0.50
Plant residue	2.04	0.44
Control	0.62	0.16

A slight decrease or increase in the atmospheric humidity may affect the crop growth and incidence of diseases in the crops. It has been reported that a condition of high humidity for several days after heavy rainfall may cause poor growth and yellowing of leaves in *kharif* (Srawan month). However, a humidity condition above average is always conducive after receding of monsoon. This impression is also reflected in the Rajasthani proverb—

> *Sawan me suryo bhalo, bhadarwa purwai*
> *Ausojan me pachua chalo, bhar-bhar gada lai*
>
> (A dry westerly wind is good in *Sawan*, but easterly good in *bhado* month, when westerly runs in *Aswin*, all these help filling the farmer's cart with food grain.)

Other water saving techniques like use of anti-transparent are very useful in arid and semi-arid areas when certain film forming chemicals are sprayed on the plant leaves which help reducing transpiration, chemicals which are found effective are folikot, kaolinite, PMA, etc. @ 1000 ppm.

Cracking of fruits is believed to be due to wide variations in soil moisture content. High air temperature was found to be the cause of cracking. Uneven soil moisture, sudden change in atmospheric temperature and humidity, long dry spell followed by heavy rainfall and hailstorm are other causes of fruit damage. Establishment of wind breaks and shelter belt prior to orchard establishment should be given due emphasis. It has been observed in some of the farmers field that where provision of wind breaks are there, abiotic stress is not a severe problem. Fruit trees like *beel*, *jamun*, *gonda*, etc. may be taken for wind break plantation to serve dual purpose.

Management of Farming System

For sustainable agriculture in watershed areas cropping systems requiring large quantity of water are always detrimental to the eco-system environment. Areas without alternative source of irrigation must be spared for hardy plants—like *ber* and *karonda* with conservation strategy including rainwater management. In agro-forestry system of *P. cineria, Ticomella undulate* and mix stand of trees; productivity of pearl millet was maximum with *P. cineria* than others (Singh, 2002).

Horti-pastoral system—It is more suitable for arid eco-system in between dunal terrains and sandy plains with annual rainfall less than 300 mm. However, this is a promising system for ensuring availability of nutritious and quality food and fodder for the masses and livestock. Here fruit trees are integrated with pasture grasses and legumes by following watershed management interventions including conservation measures. Combination *ber/karonda/aonla* + *sewan* or dhaman grass would be an ideal proposition which provide sufficient fodder for the cattle in addition to fruits for masses. Addition of organic matter to the soil is an important aspect of integrated fertility management as well as soil conservation by horti-pastoral system.

Agro-horticulture—Agro-horticulture, i.e. growing of millets or pulses with existing fruit plants is usually a sustainable system of farming in the sense that sufficient biomass is provided by the falling leaves of plants grown with the system, yet, providing production of pulses, *mateera*, *kachri* and cucumber in a young establishment of fruits as *ber/karonda*. Important combination of such farming system in arid agro system – moth bean or cluster bean + *mateera* in *ber* or *karonda* or *sanjana*. In an agri-horti system at Hyderabad several intercrops (castor, *cenchrus* and stylo) were taken between tamarind rows. The tamarind + *cenchrus* system recorded maximum net returns when compared with other combinations with stylo, castor and sole plantation (Annon., 2003).

Besides alleviating the risk, suitable fruit tree and crop combinations also generate extra income, improve productivity per unit area as a result of efficient use of natural resources and inputs and ameliorate and improve adverse agro-climate. Agri-horticultural combinations with legume intercrops such as cluster bean, cowpea and vegetables as cucurbits, brinjal and chilli could be grown as intercrops to increase cropping intensity. Atal (2005) reported that out of seven type of fruit plants grown in Bikaner region *ber* followed by *aonla* and *mosambi* recorded maximum benefits whereas pomegrenate and date palm were found to be least economical.

Diversified land use with arid fruit trees is considered the most ideal strategy for economising the productivity by harnessing the nutrients and moisture from different soil strata and nutrition security in the arid ecosystem. Such diversification calls for a proper management and utilisation of suitable combination of exotic and indigenous fruits to attain maximum returns per unit area on sustainable basis.

Planting Technique

Planting technique is an important aspect of crop production for efficient moisture utilisation in dryland areas. Time of planting should be adjusted in such a way as to assure satisfaction of nutrient and crop water requirement at that time. Obviously time of planting may be slightly advanced or delayed to avoid biotic interference like pest and diseases to sustain quality product as well as to suit the micro climate. Also aspect of *in situ* moisture conservation coupled with efficient manuring further increases the efficiency of moisture use in watershed areas. Advance planted fruit plants with use of balance nutrition has been found to be a promising technology of integrated water management in arid eco-system with quality production due to judicious combination of monsoon and tube-well irrigation, provided seedling are made available at proper time.

Paul et al. (2003) reported that plant growth hormone GA-3 recorded maximum fruit yield of cucumber var. local super green. Timely use of thio urea plant growth hormones or growth retardants as per requirement of prevailing climate and weather may prove productive.

Application of organic manure at the time of filling of pits and subsequently during annual manuring improves water storage capacity of soil. In low rainfall areas, the application of organic manure should coincide with rains. Use of manures and fertilizers at required doses and appropriate time according to age of plant is essential.

Cultural Operations

Soil moisture and proper airation are the important limiting factors in plant growth in drylands. In the arable farming the main objective of interculture is to eradicate weed plants. Moreover in dryland watershed areas for integrated watershed management in general interculture operations are accomplished to achieve triple benefits in horticulture viz., weed control, moisture conservation and root aeration. Interculture and hoeing operations increase aeration including oxygen in the growing root zone of soil and consequently help to healthy plant growth. Several estimates have indicated that more than 35 per cent moisture is saved to crop by interculture and hoeing operation in rainfed areas. For such operation to create a loose thin layer of soil mulch on soil surface blade type implements are used in the sense that blade hoe during the process of working cuts down the roots of weeds and leave the surface soil intact and loose to act as mulch. Certainly interculture operations are more desirable in dryland areas than chemical means for weed control. Manual hoeing operation are useful for water use efficiency and product quality.

Efficient Micro Irrigation

Crop water use efficiency is usually governed by method of irrigation and as such an efficient method of irrigation will go a long way in increasing water use efficiency and proper integration of water management. Traditional method of irrigation as border and check basin caused heavy loss of water in seepage and deep percolation. Sprinkler irrigation cannot be adjudged as an efficient method of irrigation for horticultural crops in arid land watershed areas. All out efforts are needed to encourage drip irrigation for integrated watershed management in this type of eco-system. In case of drip irrigation, the irrigated area may be increased by 200 per cent on account of very efficient utilisation of water with minimum losses of seepage and evaporation. Drip irrigation may be employed from rainwater storage tanks for limited area of life saving irrigation in fruit crops. In eastern part of *Thar Desert* sprinkler irrigation not only increased the area under irrigation, utilised undulating topography under cultivation but also reduced demand of agriculture labour drastically in the ratio of 3:1 compared to border irrigation. It saved 30-35% water in comparison to wasteful methods. Thus, there is need to conserve rainwater for sustainable agricultural production and development for today and to feed quality products to each and every mouth of the growing population in the 21st century. Singh (2005) reported that drip irrigation in citrus reduced irrigation water requirement by 40 per cent as compared to conventional irrigation method. He further reported that out of various micro irrigation systems the highest Water Use Efficiency observed in microjet 180° with respect to yield and quality of citrus fruits.

The response of drip irrigation in rose found effective with respect to plant growth, flowering, flower quality and cut flower yield. Plant irrigated by drip at 0.6 CPE increased plant height, number of shoots and number of flowers than surface method of irrigation.

A very efficient irrigation technology for water economy at small farms is pitcher irrigation in small fruit plants like Karonda and vegetables. In this technique of irrigation pitchers of 4-5 lit. capacity are imbedded in the ground in such a way as the mouth of pitcher remain above ground level. These pitchers are filled with water and covered by a lid to control evaporation of water. Fruit and vegetable plants near the pitcher automatically receive water from the pitcher by capillary action. Care should be taken to refill the pitchers as soon as half of the pitcher get empty.

REFERENCES

Annonymous (1996). Res. highlights, DOR, RAU, Bikaner, p. 23.

Annonymous (2003). Annual Report, CRIDA.

Atal, R.P. (2005). Personal Communication from Atal Krishi Farm (Bikaner).

Paroda, R.S. (2000). The Hindu—Survey of Indian Agriculture.

Paul, B.B., Khajuria, V. and Prasad, V.M. (2003). *Effect of GA-3 on Yield of Cucumber Varieties*. National Symposium on Resource Management for Eco-Friendly Production. Feb. 2003 (I.S.A.).

Singh, G.D. (2002). *Watershed Management Technology for Hot Arid Regions*. Biennial Workshop of N.A.T.P. (ICAR) Project, H.A.U., Hisar (July).

Singh, S. (2005). *I.C.A.R. News*, II(1): 2005.

Participatory Irrigation Management and its Success in Tamil Nadu

11

M. Shantha Sheela, R. Rajasekaran and *K. Palanisami*

Introduction

Since inception of community development programmes and also various agricultural development programmes from 1952 onwards, there has been an emphasis on transferring the technology to the farming community through the government agencies. But involvement of local people, i.e. farmers, farmwomen, farm youth, local legitimisers and village-level peer groups was not much. That is why the tempo of agricultural and rural development could not catch up to the tune expected despite heavy resources and energy spent on achieving the objectives. The desire for the change did not come from within the heart of the rural people. They adopted the improved technologies just because some amount of help/subsidy/credits/ inputs was supplied to them either on subsidised basis or free of cost. Further, the responsibility for mobilising people at the gross-root level also rested with government machinery rather than the local people. Thus, the local people were not bound to fully implement the programmes and to see that these programmes are sustained and they result into tangible gains. Now, it has been widely realised that organisations engaged in research and extension must have an effective feedback system from the users of the technology in order to make the research and extension system demand-driven and market-driven. Thus, the crucial factor in improving the feedback system is to organise farmers into functional groups. These groups are virtually acting as secondary disseminators of the improved water management technologies among the rest of the farmers. So the results are fruitful, only if the transfer of technology that involves farmers from assessing the project to implementation stage.

Therefore, any improvement in on-farm water management will conserve water that can be used to expand more intensive agriculture that will increase farm income. The performance of monsoon rain will decide the water resources of Tamil Nadu. Among the area irrigated under different sources, significant reduction observed under tank irrigation over years. The share has declined from 28 to 24 per cent during this 20 year period. Introduction of proper water management strategies on the farm greatly enhance the percentage of land cultivation. Improving the on-farm water application efficiencies can increase the command area supported by a single tank. In the case of well irrigation, the major issue is the over-exploitation of the aquifers. Number of wells in the state has increased from 60,000 in 1960s to 1.8 million in 1990s and about 1.4 million wells are energised. The average area irrigated per well has declined from 1.5 ha from 1960s to 0.6 ha in 1990s. The cost of well investment has also increased from Rs. 30,000 to Rs. 1.5 lakh during the above period. The average failure rate of tube-well is about 67 per cent (Palanisami et al., 1997). The externalities to the small and marginal farmers are so high that the well abandonment is also increasing over years. Improved water management strategies in this area not only save the losses occurring due to surface irrigation, it also reduce the over-exploitation of wells and borewells. Among the districts water scarcity is high in Coimbatore district and the area irrigated by wells, canals and tank was 1,34,223 ha (72.32%), 48,675 ha (26.23%) and 2,696 ha (1.45%) respectively. Even though, there are three sources of irrigation which are not able to meet the irrigation demand of farmers in Coimbatore district. Water Technology Centre at Tamil Nadu Agricultural University involves few technology transfer projects. All these projects it involves people from the identification of the location, technology suited to the crop and field to field trails. That's why the transferred technology is viable and sustainable.

Objectives

(a) Farmers' discussion and Participatory Irrigation Management (PIM);

(b) Water and soil resources;

(c) Implementation of improved irrigation practices;

(d) Impact of the improved irrigation practices; and

(e) Facilitate the transfer of technology by conducting field day.

Methodology

Selection of study area

Agragarasamakulam village in Coimbatore district of Tamil Nadu State was selected for the study. Achieving this task was found to be easier in this district in the early stage of the study itself, because of their familiarity with Tamil Nadu Agricultural University and the farmers' progressiveness towards the use of new concepts and methods in farming.

Selection of respondents

The introductory meeting of the project "Implementation of System Improvement and Water Management Strategies in Agragarasamakulam Tank in Coimbatore District" was conducted at Agragarasamakulam village. All farmers in the villages were invited with the help of village panchayat presidents to attend the meeting. The interested farmers attended the meeting. Discussion with scientists and farmers were arranged.

After that, the scientists who are involved in the project visited the farmers' field and conducted a preference survey to identify which type of irrigation technology they prefer to adopt in their field. Based on water availability, crops grown and present water management practices, the fields were selected. Those farmers who were accepted to give their co-operation for making demonstration plots in their farm and to take care of the demonstration plots were selected.

The selected farmers were taken to Tamil Nadu Agricultural University—Water Technology Centre—irrigation cafeteria to know about the various irrigation technologies demonstrated to different crops and also to farmers' fields where these technologies were actually implemented by the farmers and finally nine farmers were selected.

Technology	Farmers preferred	Farmers selected
Drip irrigation	25	6
Drip fertigation	10	3

Identification of irrigation management strategies based on farmers' need

To identify the cost-effective irrigation method the details of crop and interest of farmers were collected and field visits were made. Most of the farmers prefer drip and drip fertigation systems. Drip system is economical, if the design provides optimum size of main, sub-main and laterals. Discharge and spacing is important to give the required quantum of water and wetted area for the crop. Besides proper design, economical lay-out is important in deciding the cost of the systems for annual and close spaced crops such as banana, turmeric, curry leaf and vegetables, etc. since the cost of laterals and emitters decide the cost of the system. So it is important to reduce the length of laterals and number of drippers required per unit area by properly adjusting the crop geometry, without sacrificing the population of the crop. Based on the crop spacing the layout such as single line drip, alternate row (in line) drip and paired row drip is decided. Fertigation is advantageous along with drip irrigation to save fertilizer and labour and also yield and quality improvement. Venturi assembly is used for fertigation in the drip system.

Findings

Water and soil resources

Soil and water samples from the selected farmers' fields were collected and analysed for their nutritional status and quality.

The selected farmers were supplied with a Tamil Nadu Agricultural University—Soil Health card for maintaining and updating their soil health status periodically.

In this village the farmers are using only the groundwater for cultivation of the crops. The existing crops are irrigated with well water with a depth of more than 100 feet in open wells and 650-750 feet in borewells in all farms. The crops are under-irrigated because of the poor rainfall and deficit water supply. Water budgeting was worked out for the selected crops in the farmers field and alternative strategies were suggested for successful farming with the available water.

The worked out water budget is given in Table 1 as an example for one farmer. Similarly, the water resources of the selected farmers were worked out and presented in Table 2.

Impact of the Improved Irrigation Practices

Drip irrigation in sugarcane

The farmer was happy and equated the water requirement of 1.25 acre of sugarcane in surface irrigation to three acre under drip irrigation. The results of the study indicated about 40 per cent water saving with 25 per cent increase in production. A comparison of the drip irrigation with surface irrigation in sugarcane is presented in Table 3.

Drip fertigation in banana

The farm has water saving of 40% in drip irrigation. Since the water is delivered exactly at the root zone of the crop there is no possibility of weeds grown in other areas which considerably reduces the labour on weed management. The fertilizers are also mixed with water and readily available to the crop and also matching the requirement of the crop. The farmer was happy with a yield of 28.75 t ha^{-1} as against 20 t ha^{-1} through conventional irrigation practices, which is 44% higher yield. Now the farmer is convinced and wanted to lay drip irrigation system to vegetable crops also. The benefits realized because of the irrigation system are presented in Table 4.

Drip irrigation in mulberry

Mr. Uthayakumar farmer with a well-water of high EC, Ca, Mg and Na salts and not amenable for surface irrigation because of the fear that it will pollute the entire soil column was approached for improved water management strategies in mulberry. Drip irrigation in mulberry was suggested to the farmer. The farmer is now happy with a strong feedback that he reduced the water application to 1/10th of the water he applied

Table 1: Present Status of Water Use and Water Budgeting to Sugarcane for 4 Acres (m^3)

Sl. No	Particulars	Jan	Feb	Mar	Apr	May	June	July	Aug	Sept	Oct	Nov	Dec	Total
1.	Water currently available to the crop	2601	1762	1951	1888	1625	1573	1300	1300	1888	2601	3147	3251	24892
Effective rainfall		78	85	135	871	498	359	943	590	867	2352	2420	500	9698
2.	Actual water requirement of sugarcane	2400	2400	2400	2400	2400	2400	1600	1600	1600	2880	2400	2400	26880
Irrigation requirement		2322	2315	2265	1529	1902	2041	657	1010	733	528	–20	1900	17182
Excess/Deficit		279	–553	–314	359	–277	–468	643	290	1155	2073	3167	1351	7705
3.	Irrigation requirement if drip is used to sugarcane	1600	1600	1600	1600	1600	1600	1120	1120	1120	1760	1600	1600	17920
Excess/Deficit		1079	247	486	1159	523	332	1123	770	1635	3193	3967	2151	16670

Table 2: Water Resources of Selected Farmers

Sl. No.	Area (ha)	Crop	Total water available / year (m^3)	Rainfall (m^3 / year)	Actual req. of the crop (m^3)	Total water availability / year	
						Excess months	Deficit months
1	1.6	Sugarcane	21388	12426	26880	9	3
2	3.2	Sugarcane	44640	24852	53440	9	3
3	2.0	Sugarcane	43466	15532	34320	10	2
4	4.0	Coconut	29478	31065	28800	12	0
5	1.6	Sugarcane	30672	12426	26880	11	1
6	2.3	Sugarcane	22355	17940	38577	5	7
7	1.6	Sugarcane	24892	12426	26880	8	4
8	8.0	Coconut	35916	62130	57600	8	4
9	5.0	Mulberry	22210	34320	57360	9	3

Table 3: Comparison of Drip with Surface Irrigation in Sugarcane

Sl. No.	Items	Surface irrigation	Drip irrigation
1.	Annualised drip system cost (Rs ha^{-1})	—	10,000
2.	Cost of cultivation (Rs ha^{-1})	60,250	54,125
	Total	60,250	64,125
3.	Gross income (Rs ha^{-1})	1,02,500	1,33,250
4.	Profit (Rs ha^{-1})	42,250	69,125
5.	Cost Benefit ratio	1.70	2.46
6.	Water used (mm)	1900	1140
7.	Water saving (mm)	—	760 (40%)
8.	Yield (t ha^{-1})	117	147
9.	Water use efficiency (kg ha^{-1} mm^{-1})	61	128

Table 4: Comparison of Drip Fertigation with Surface Irrigation in Banana

Sl. No.	Items	Surface irrigation	Drip fertigation
1.	Annualised drip system cost (Rs ha^{-1})	—	10,000
2.	Cost of cultivation (Rs ha^{-1})	53,080	52,230
	Total	53,080	62,230
3.	Gross income (Rs ha^{-1})	1,40,000	1,75,000
4.	Profit (Rs ha^{-1})	86,920	1,12,770
5.	Cost Benefit ratio	2.63	2.81
6.	Water used (mm)	1900	1140
7.	Water saving (mm)	—	760 (40%)
8.	Yield (t ha^{-1})	20	28.75
9.	Water use efficiency (kg ha^{-1} mm^{-1})	10	25
10.	Fertilizer use efficiency	14.82	21.92

earlier by surface irrigation. The other interesting features were control of weeds, restricted fertilizer application zone, alleviation of micro nutrient deficiency and improvement in the quality of the leaf. A comparison of the drip irrigation with the surface irrigation is given in Table 5. In view of his foresight in optimally utilising the saline water for the area, adopting micro irrigation technology, realising the impacts of drip fertigation and integrated farming technologies, transferring the technologies to the farmers in that region and the acceptance of this method by the farming community at large he was awarded "The Jain INCID Sookshma Sinchai Puraskar Award".

Table 5: Comparison of Drip Fertigation with Surface Irrigation in Mulberry

Sl. No.	Items	Surface irrigation	Drip fertigation
1.	Annualised drip system cost (Rs ha^{-1})	—	10,000
2.	Cost of cultivation (Rs ha^{-1})	49,325	45,075
	Total	49,325	55,075
3.	Gross income (Rs ha^{-1})	72,000	86,400
4.	Profit (Rs ha^{-1})	22,675	26,325
5.	Cost Benefit ratio	1.45	1.56
6.	Water used (mm)	1800	1080
7.	Water saving (mm)	—	720 (40%)
8.	Leaf yield (t ha^{-1})	15	18.25
9.	Water use efficiency (kg ha^{-1} mm^{-1})	8.33	16.89

Drip Irrigation for Coconut

The farmer expressed the water saving of 52%, i.e. 96 liters of water is applied daily through four drippers of 8 LPH for three hours than 200 liters $tree^{-1}$ day^{-1} in the conventional surface irrigation. The yield of nuts were around 80 $tree^{-1}$ $year^{-1}$. Now he is expanding the drip system for the entire 8 acre of coconut. The comparison is given in Table 6.

It could be observed from the Table 7 that the selected beneficiaries had pointed out an overall impact of water saving and increased yield obtained as a result of improved method of irrigation. The water saving could be attributed because of no water loss in the field and conveyance. The farmers had a very positive notion that the scientists are with them for any technical know-how for their implementation and maintenance which

Table 6: Comparison of Drip Fertigation with Surface Irrigation in Coconut

Sl. No.	Items	Surface irrigation	Drip fertigation
1.	Annualised drip system cost (Rs ha^{-1})	—	2,900
2.	Cost of cultivation (Rs ha^{-1})	7,520	6,270
	Total	7,520	9,170
3.	Gross income (Rs ha^{-1})	42,336	55,776
4.	Profit (Rs ha^{-1})	34,816	46,606
5.	Water used (lit $tree^{-1}$)	200	75
6.	Water saving (mm)	—	125
7.	Yield (Nuts ha^{-1})	10,080	13,440

Table 7: Garret Ranking on Overall Impact, Increased Yield, Water Sàving and Perceptions due to the Implementation of Improved Irrigation Practices

Sl. No.	Overall impact	Garret score	Rank
1	Increased water saving	100	1
2	Increased yield	88	2
3	Labour saving	66	3
4	Increased awareness	55	4
5	Increased adoption	44	5
	Increased yield		
1	Improved method of irrigation	99	1
2	Fertigation	77	2
3	Stress alleviation	55	3
4	Weed control	44	4
5	Easy operation	33	5
	Water saving		
1	No water loss in the field	100	1
2	No conveyance loss	88	2
3	Water applied in root zone	77	3
4	Timely irrigation	66	4
5	No weeds	55	5
	Perception		
1	Scientific personnel at their reach	100	1
2	Technology is user friendly	77	2
3	Problems alleviated with technical know-how	66	3
4	Easy maintenance	55	4
5	Suitable for high value crops	44	5

created a lot of confidence. It is expected that other farmers within the village as well as in the neighbouring villages will be able to adopt these technologies on their own investment.

Field day

Finally, field day was conducted to disseminate the results of the demonstration. All the beneficiaries of the project expressed their happiness and the results achieved in their field. Those who attended the meeting are willing to implement the drip system, drip fertigation in their field of their own cost.

Conclusion

Any research is fruitful if it is transferred effectively to the needed people. For transferring the technology different extension strategy is needed. From this research we could realise that still personal contact is the most effective tool in the Transfer of Technology. Combined with this the meetings, discussions, small tours, result demonstration and field day gave a great success in the transfer of micro-irrigation technologies.

REFERENCE

Palanisami, K., Balasubramanian, R. and Mohamed Ali, A. (1997). *Present Status and Future Strategies of Tank Irrigation in Tamil Nadu*, Publication of Water Technology Centre, Tamil Nadu Agricultural University, Coimbatore.

Social Gains and Effect of Encroachment in Tank Irrigation System

12

D. Karpagam and *C. Karthikeyan*

Introduction

Irrigation has played a very important role in India's agriculture. Tanks play a dominant role in irrigation in the following Indian states like Andhra Pradesh, Karnataka, Kerala, Madhya Pradesh, Maharashtra, Tamil Nadu and Uttar Pradesh where tank irrigation accounts for an appreciable area. According to Thinagaran (1997) during the year 1992-93 there were 7,176 tanks with ayacuts of 40 ha or more and 31,687 tanks with command area of less than 40 ha. About 6,28,830 ha have been irrigated by tanks in Tamil Nadu as against about 8,52,000 ha by canals and 1,200,100 ha by wells.

Encroachment of tank was observed in different parts namely supply channel, catchment area, foreshore area, tank bed, etc. Due to the encroachment of the lands by these, water filling in the tank gets decreased which eventually deter the availability of water to the farmers, dwelling alongside the command area. Apart from this, the encroachers in the foreshore area break open the tank bunds thus avoiding the water to fill at full tank level. These facts remained as the cause for the social conflicts between encroachers and farmers in the tank command (Palanisami et al., 2001).

Ramakrishnan (2000) reported that formation of sand dunes and encroachments by individuals have almost kept tanks at bay. Some of the tanks had shrunken as a result of encroachment. The present estimate of loss caused to India as a result of encroachment and other factors was put at Rs. 5000 crore.

Encroachment had direct influence over the tank performance. To improve the performance of tanks eviction of encroachment should be thought of seriously. In this context, studying the consequences of encroachment is important. Hence, the present study has been designed to find out the consequences of encroachment and provide suggestion to evict the encroachment.

Materials and Methods

This study was conducted in Theni district of Tamil Nadu. Karuvelankulam tank of Kulaiyanur village and Bodinayakkanur taluk was selected as a case for conducting the case study on encroachment in tank irrigation system based on the tank which had the problem of encroachment and which was depended upon by farmer, solely for irrigating their crops. For studying the perception of command area farmers towards the consequences of encroachment in tank irrigation system, 24 farmers were selected randomly from the ends of the tank viz., Head, Middle and Tail. Hence, a total of 72 farmers comprised the sample size of farmers for the study. Semi-structured interview schedule was developed to enhance the validity and reliability of the information, PRA techniques such as group discussion and consequences diagram were used. Percentage analysis was used for making simple comparison.

Results and Discussion

Karuvelankulam tank

The Karuvelankulam tank is located in Kulaiyanoor village of Uthamapalayam revenue division, Bodinaickanur taluk, Bodinaickanur block and Kulaiyanoor panchayat. It is a system tank which received water from the Kottoor dam. The tank also received water from 'Mullingasamy odai' which was divided into three streams namely Ellodai, Muthuirrulappa Pillai odai and Ramar odai. The length of the tank bund is 875 m. The registered ayacut of the tank is 99 ac. The free catchment area is 2.59 sq.m. There were two sluices in the tank known as 'Chinnamadai' and 'Periyamadai'. Of the 99 ac, Chinnamadai covered 35 ac and Periyamadai 64 ac of command area.

History of encroachment

During 1972, Mr. Mottaiyan started the encroachment in tank. Mr. Mottaiyan family basically landless who leased in by heirs the tank for cattle manure and green manure, as days rolled started to encroach the area. The adjacent land owners to the foreshore area started to extend their encroached area. The villagers next to these villages who owned agricultural land in these villages started to have an eye on those encroached area. Villagers who dug out silt, started cultivation in the same area of the tank. Before 1980, only crops like sorghum, cumbu, castor, water melon, cucurbits and trees like silk cotton and coconut were grown. But during 1980, Mr. Ponnangan started paddy cultivation using his well-water influenced other farmers also to practice paddy cultivation. Farmers who cut the fodder grasses encroached the same land for their own purpose. Some farmers encroached very small area not for the purpose of cultivation but with a view of selling it to other large encroachers.

Encroacher's background

Out of 24 ac of total area of the tank, 22.65 ac was encroached by 25 persons. Among the 25 encroachers two encroachers occupied outside the tank bund area for coconut cultivation. Others encroached the foreshore area and water spread area of the tank. Among those encroachers only one encroacher belonged to Scheduled Caste and others belonged to Thevar community. Thevar community was dominant in this village that leads to ultimate Thevar population over there. Among 25 encroachers, 7 were adjacent field owners, who had their 'patta' land near foreshore area of the tank, 10 were landless and others were farmers of the command area and garden and dryland of the same village.

Consequences of the Encroachment

The consequences of encroachment in tank irrigation system as perceived by the command area farmers are presented in Table 1.

Table 1: Consequences of Encroachment in Tank Irrigation System

(n=72)

S. No.	Consequences of encroachment	No.*	%
I.	**CONSEQUENCES IN THE SOCIETY**		
(a)	**Positive Consequences**		
•	**Socio-economic consequences**		
(i)	Co-operation of command area farmers for formation of TFA	36	50.0
(ii)	Betterment of livelihood of encroachers	30	41.6
(iii)	Government get revenue through assessment	25	34.7
(iv)	Employment opportunity to agricultural labourers	20	27.7
•	**Ecological consequences**		
(i)	Maintain the hygienic condition of the tank	13	18.0
(b)	**Negative Consequences**		
•	**Socio-economic consequences**		
(i)	Conflicts between encroachers and command area farmers	69	95.8
(ii)	Less revenue from tank	35	48.6
(iii)	Migration of people	6	8.3
•	**Ecological consequences**		
(i)	Fish culture affected	40	55.5
(ii)	Reduction in animal population	12	16.6
II.	**CONSEQUENCES ON AGRICULTURE**		
(a)	**Positive Consequences**		
•	**Socio-economic consequences**		
(i)	Higher yield from encroached land	5	6.9
•	**Ecological consequences**		
(i)	Seasonal and permanent cultivation in encroached land	72	100.0
(b)	**Negative Consequences**		
•	**Socio-economic consequences**		
(i)	Reduction in water spread area of the tank	72	100.0
(ii)	Reduction in storage capacity of the tank	50	69.4
(iii)	Yield reduction	47	65.2
(iv)	Reduction in command area	41	56.9
(v)	Problems in silt collection	36	50.0
(vi)	No rainwater harvesting	36	50.0
(vii)	High cost of cultivation due to purchase of water	25	34.7
•	**Ecological consequences**		
(i)	No tank filling	36	50.0
(ii)	Changes in cropping pattern	30	41.6
(iii)	Crop failure	25	34.7
(iv)	Groundwater depletion	15	20.8
(v)	Flooding	15	20.8

* Multiple response recorded

Consequences of Encroachment in the Society

Both positive and negative consequences occurred in the society due to encroachment in the tank irrigation system are given below.

(a) Positive consequences

It could be seen from the Table 1 that 50 per cent of the command area farmers perceived that there was remarkable co-operation among the command area farmers for the formation of TFA to evict the encroachment. More than one-third (41.6%) of the farmers perceived the betterment of livelihood of encroachers and 34.7 per cent of the farmers perceived that government got revenue through collection of assessment charges from encroachers. Encroachment in tank irrigation was done for agriculture, so there was better employment opportunity to agricultural labourers. This was perceived by 27.7 per cent of the command area farmers. The hygienic condition of the tank was maintained and it was attributed to the encroachment. This fact was perceived by 18 per cent of the command area farmers.

(b) Negative consequences

There were conflicts among command area farmers and between the encroachers vs. command area farmers because of encroachment in the tank. This was perceived by 95.8 per cent of the farmers. Fish culture was greatly affected by the encroachment and was perceived by slightly higher than half (55.5%) of the command area farmers. Nearly half of the farmers (48.6%) perceived that due to encroachment, there was reduction in revenue from tank. Other consequences such as reduction in animal population due to decreased water storage and increased migration of people to cities or towns to earn their livelihood were perceived by 16.6 per cent and 8.3 per cent respectively.

Consequences of Encroachment on Agriculture

(a) Positive consequences

It would be observed from Table 1 that all the command area farmers perceived that there was seasonal and permanent cultivation in encroached land. Higher yield could be obtained from encroached land was perceived by 6.9 per cent of the command area farmers as one of the positive effect.

(b) Negative consequences

The major negative consequence due to encroachment was reduction in the water spread area of the tank. It was perceived by all the command area farmers. Storage capacity of the tank was reduced and it was perceived by 69.4 per cent of the farmers. About 65 per cent of the farmers understood that due to encroachment crop yield was greatly reduced. More than half of the command area farmers (56.9%) perceived that due to encroachment in tank, command area of the tank was reduced. Half of the command area farmers (50%) perceived that encroachment of tank area resulted in negative consequences like problem in silt collection (i.e. encroachers prevented the silt collection), absence of rain water harvesting and tank filling.

Conventional cropping pattern was changed due to encroachment. This was perceived by 41.6 per cent of command area farmers. More than one-third (34.7%) of command area farmers perceived that cost of cultivation of paddy crop was increased due to the purchase of well water from well owners as well as due to crop failure in command area, as a result of the encroachment in the tank system. Slightly more than one-fifth of the command area farmers (20.8%) perceived that due to encroachment there were flooding in command area due to breakage of sluice by encroachers deliberately and depletion of groundwater in the area.

The reason attributed that encroachers grabbed the tank water and irrigated their encroached land. As a result, command area farmers did not receive sufficient water. So at that point of time, conflicts occurred between command area farmers and encroachers. Again within encroachers, conflicts occurred for extending their encroached land. Encroachment was done for doing agriculture. So encroachers levelled their encroached land in tank bed and formed bund around their encroached land, because of this the water spread area of the tank got reduced. It caused the reduction of storage capacity of the tank.

Consequence Diagram

Due to encroachment in the tank system both positive and negative consequences were perceived by the command area farmers. The consequence diagram as prepared by the key informants and respondents is shown in Fig. 1.

Positive consequences

- Encroachment in tank system would induce the command area farmers to form the Tank Farmers Association (TFA) for eviction of encroachment. By the formation of TFA the cooperation among the command area farmers was increased. Through this way the encroachers in the tank system were evicted successfully.
- Encroachers were mostly landless agricultural labourers. They knew only farming and they got employed only a part of year. So their income was too low ranging from Rs. 10,000 to Rs. 15,000. By the way of encroachment, they got additional income of Rs. 20,000 to Rs. 40,000 for leading their life.
- Because of the cultivation of crops like paddy in the encroached area, it provided employment opportunities to agricultural labourers.
- All the encroachers in the tank paid assessment charges depending on their extent of encroachment. So, government got additional revenue due to encroachment.

Negative consequences

- Due to the encroachment of the lands by these, water filling in the tank gets decreased which eventually deter the availability of water to the farmers, dwelling alongside the command area. Apart from this, the encroachers in the foreshore area break open the tank bunds thus avoiding the water to fill at full tank level. These facts remained as the cause for the social conflicts between encroachers and farmers in the tank command

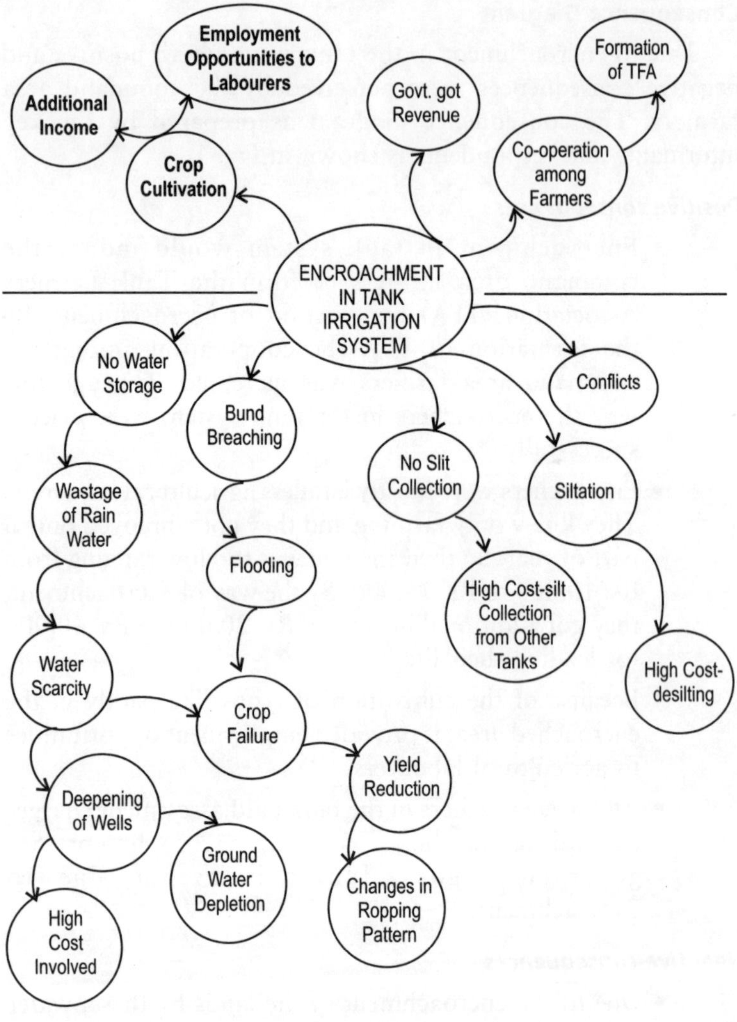

Fig. 1: Consequences of encroachment in Karuvelankulam tank

- Due to encroachment in tank bed, farmers could not collect silt from the tank which had to be added to soil for improving soil fertility. Encroachers prevented the farmers to collect silt. So farmers had to make arrangements to collect silt from other tanks. This involved more cost which farmers cannot afford.

- Due to encroachment in the command area, first season paddy crop was only successful. During second season paddy crop, they got water up to half of the crop period. So nearly for 8-10 irrigations, they got water from well owners. Few farmers, brought well water from garden land to wet land through underground pipes, minimum 2 km distance at the rate of Rs. 2 lakh. So cost of cultivation increased due to purchase of well water at the rate of Rs. 1000/season. Because of exploitation of well water, there was depletion of groundwater.
- Encroachment and less rainfall led to scarcity of water and they could not cultivate paddy even for one season. Since they got water only for half of the crop period, yield was low. So they changed cropping pattern and cultivated crops like sorghum, groundnut and cotton as rainfed crops.
- Feeder canal namely 'Ellodai' joins the tank near surplus weir. There was encroachment near surplus weir. Due to this, encroachers did not allow water from Ellodai into the tank as they wanted to protect their encroached land. So that water directly went to 'Suruli' river. So there was no water harvesting.
- During heavy downpour of rainfall and high water supply period, encroachers to safeguard their encroached land from submergence opened the sluices which resulted in flooding in the command area. As a result, crop cultivation was affected and yield was reduced.

Suggestions for Eviction of Encroachment

An attempt has been made to identify overall suggestions expressed by command area farmers of Karuvelankulam tank for eviction of encroachment. The results are presented in Table 2.

Table 2: Suggestions for Eviction of Encroachment

(n=72)

S. No.	Suggestions	No.	%
1.	Properly survey the tank, placing boundary stones around the tank	72	100
2.	Impartial and bold leadership is a must for evicting the encroachment	65	90.27
3.	Villagers should have unity to evict the encroachers	55	76.36
4.	Government should bring the tank maintenance work under farmer's control	45	62.50
5.	Eviction should be done without any discrimination	43	59.72
6.	Farmers should realise the ill-effects and take steps for evicting the encroachment legally	15	20.83

It is evident from Table 2 that majority of the command area farmers suggested proper surveying of the tank and fixing boundary stones around the tank (100%), impartial and bold leadership by capable person is a must for evicting the encroachment (90.27%), villagers should have unity to evict the encroachers (76.36%), government should handover the tank under farmer's control (62.5%), eviction should be done without any discrimination (59.72%) and farmers should realise the ill-effects and take steps for evicting the encroachment legally.

Conclusions

Tanks being a common property resource had uniformly suffered from encroachment problem to varying degrees in the state. Now a days, there is a need for strengthening the water harvesting structures. The stored water could be used for domestic and agricultural purposes. But almost all the tanks have been encroached for various purposes. Encroachment had direct influence over the tank performance. To improve the performance of tanks eviction of encroachment should be thought seriously by users and tank management authorities.

REFERENCES

Palanisami, K., P. Paramasivam, C. Karthikeyan and A. Rajagopal (2001). "Sustainability of Tank Irrigation Systems in South India." Water Technology Centre, TNAU, Coimbatore.

Ramakrishnan, D. (2000). Tanks and Grass Lands. *Tank Cascade—A Development Journal on Irrigation Tanks in South India*, DHAN Foundation, Madurai, Vol. 2, Sept. & Dec., 2000, No. 3 and 4, pp. 94-98.

Thinagaran, M. (1997). "Tanks and Social Forestry—The Role of Forest Department." National Seminar on Farmer's Participation in Tank Rehabilitation and Management at Irrigation Management Training Institute, Trichy, Tamil Nadu, Jan. 24-25.

Need for Integrated Watershed Development Programme

13

E. Mahalakshmi Geetha, N. Kalaichelvi
and *A. Manickavalli*

Introduction

Water is the most limiting factor for achieving sustained agricultural production. But the demand of this scarce input is increasing continuously. Increasing area under irrigation and the use of non-conventional sources of water for irrigation are not economical. Modern agriculture should be exploited to achieve the twin objectives of higher productivity and better water use efficiency. More than 50 per cent of the extractable groundwater has been tapped in our country. In Tamil Nadu, about 70 per cent of the groundwater has already been tapped. As the groundwater sources cannot be exploited further and at the same time as the demand for the available water resources are in steady increase, the available water resources should be used carefully. Today, industries and urban population have increased their demands spectacularly on the available water and their demands would have to be met at the cost of agriculture.

Why Water Scarcity?

- Low annual rainfall and its uneven distribution
- Monsoon failures
- Polluted available water resources
- Improper and over-exploitation of surface and groundwater
- Increasing population
- Reduction in forest areas

- Improper cultivation practices
- Indiscriminate sand quarrying in the riverbeds.

What is Watershed Management?

Watershed management is an integration of technologies within the boundaries of drainage areas for optimum development of land, water, plant resources to meet the basis and minimum needs of the people living in that area (watershed) on sustained basis. The factors associated with watershed are geology, soils size, shape, topography, drainage, land cover, rainfall, and socio-economic conditions of the people. As the problems and the needs of treatments of watersheds tend to vary from place to place, the planning and activities also tend to vary depending on the location specific aspects. For example, a possible micro watershed can be identified or selected and with the help/cooperation or involvement of the people living in and around the area, forest development/regeneration and watershed development works could be taken up. This project would conceptualise a unique blend of forest development, watershed development and rural development. This three-pronged strategy would give equal focus on regeneration of forests, hydrological improvement to downstream lands and poverty alleviation and community development to uplift the socio-economic condition of the interface villages.

Development Programmes

The Government of India through various departments like, department of land resources, department of agriculture, department of environmental and forest, department of rural development, NABARD, etc. has launched the following watershed development programmes:

1. Integrated Wasteland Development Programmes (IWDP)
2. Comprehensive Watershed Development Programme (CWDP)
3. Drought Prone Areas Programme (DPAP)
4. Desert Development Programme (DDP)

5. Western Ghat Development Programme (WGDP)
6. Hill Area Development Programme (HADP) and
7. National Watershed Development Programme (NAWDPRA).

Apart from this, external funding agencies like Ford Foundation, DANIDA, SIDA, etc. are also financing these development programmes (in agricultural and forest watersheds). The main objective of these development programmes was to use the land and water resources to get the maximum benefit with involvement of the people living in the watershed. The activities involved in these programmes are:

1. Development of land and water resources
2. Development of agriculture and forest area and
3. Developing the livelihood of the rural people including the landless and labourers.

In our country, a huge amount is spent every year by various agencies. But, it is not implemented on an integrated manner by all the departments on a watershed basis in a saturated way. Still, many people in the top administration who have a say in approving the watershed development programmes are not convinced about the impact and benefit of these programmes. In the watershed/wasteland/forest development programmes a major amount is allotted to land and water resources development especially for recharging groundwater.

The impact studies carried out by our scientists has indicated the following:

(a) The results of the study conducted by Water Technology Centre of Tamil Nadu Agricultural University have indicated that construction of water conservation measures like percolation pond, check dams, farm ponds and renovation of tanks increase the storage capacity in the watershed to store the excess, run-off, has enhanced which will improve groundwater recharge. In Coimbatore district, the data revealed that the water levels in the wells have increased from 2′ to 14′. Also, the farmers have reported that these measures have reduced the water scarcity during most part of the year.

(b) The study conducted in various watersheds by Soil and Water Conservation Research & Training Institute, Ooty (ICAR) has demonstrated the use of simple and surrogate measures such as water table rise, increasing perenniality in wells, increasing recuperation, increasing irrigation and crop diversification.

(c) Studies conducted by the watershed development project in Karnataka stated that there was a direct benefit of improvement in the groundwater level, which reduces much of the burden of the government.

(d) Present emphasis on groundwater depletion needs the strategies for artificial recharge for aquifers. Soil and water conservation measures adopted on watershed basis have definite impact on replenishment of groundwater. This has been proved by the results obtained from the several watershed management projects of Dehradun.

(e) A study conducted to observe the effect of contour bunding in groundwater recharge in Theni district of Tamil Nadu revealed an increase of about 14 per cent in the water yield of wells from bunding.

(f) The minor irrigation department of Andhra Pradesh studied the impact of percolation tanks and check dams. It was stated that these structures were very useful in drought affected areas. Also, the farmers are able to reap the benefits from the immediately subsequent years.

The percolation ponds are quite effective in augmenting groundwater. A number of wells have been benefited by recharge under percolation ponds and the water level has been considerably raised due to construction of percolation ponds.

Conclusion

It is clear from the above observations that the impact of watershed development works like construction of percolation ponds, check dams, soil conservation measures like contour bunding, compartmental bunding, vegetative or stone barriers will help in augmenting groundwater in the watershed and the farmers in the watershed will raise crops in more areas under irrigation. This is very important for our country as farmers in India mainly depend on the monsoon, which is erratic, uneven and fails once in 3-4 years. The watershed development work will help in storing and recharging the groundwater. When the rainfall is more the water is stored/conserved and this water could be used in the subsequent drought years. Therefore, it is needless to say that much attention should be given to take up integrated or comprehensive watershed development programmes in the country to mitigate water scarcity.

REFERENCES

"Drip Irrigation for Sustainable Water Management", by R. Veeraputhiran, in Kissan World, January 2005, Vol. 32, No. 1.

"Impact Assessment of Watershed Development Works in Ground Water Recharge", by Dr. R.K. Shivanappan, in Kissan World, March 2005, Vol. 32, No. 3.

"Water Harvesting in Farm and House", by P. Bhuvaneswari, in Kissan World, June 2005, Vol. 32, No. 6.

Paired Comparison Ranking for Participatory Field Problem Identification

14

V.K. Jayaraghavendra Rao and *N. Sandhya Shenoy*

Participatory Rural Appraisal (PRA) practitioners in the field of agricultural research adopt RBQ (Rank Based Quotient methodology) on which the Value Based Index (VBI) is calculated for field problem identification and prioritisation. Not only is this procedure time consuming but difficult to administer in field situations because a large number of respondents have to be contacted and the problems ranked and prioritised. The limitation of the human brain to rank a maximum of six variables at a time further puts binding on the RBQ methodology when more than six problems have to be ranked and prioritised. Hitherto the PRA training for both the foundation course and the senior level participants the RBQ methodology was used for field problem identification, but during the recent senior level programme on PRA/PLA (Participatory Learning Action) for research and extension in agriculture between September 22-29, 2005 the paired comparison ranking technique to identify field problems was done as a deviation from the RBQ. The advantage of this in participatory field situations is saving of time and avoidance of drudgery by the researcher in administering the conventional RBQ vis-à-vis paired comparison technique.

Methodology

The RBQ methodology consists of the formula for the Rank Based Quotient:

$$RBQ = \frac{\sum f_i(n+1-i) \times 100}{N \times n}$$

where,

i = Concerned ranks (1 to n)

N = Total number of farmers (30-50 normally)

n = Numbers of rank

f_i = Frequency (number of farmers reporting that particular problem)

Once the RBQ are found out for each problem, the top most researchable issue in the village is identified by calculating the value based index (VBI) for each RBQ.

VBI = RBQ × Average loss experienced × Area of crops/ animals

The problem with the maximum value based index is identified as the top most researchable problem.

The Paired comparison methodology

In this case the problems experienced by the farmers are listed on separate cards using a mapping exercise on the floor by rangoli, the problems two at a time are ranked as higher and lower by a group of 50 farmers in our case who did the exercise at Survel village, Narayanpur Mandal on 26-27th September 2005 by raising their hands and frequencies noted. Their responses were tabulated and final ranking table on 12 problems was made. Priority ranking matrix of the responses were made. The criteriae were noted for justification of problem prioritisation. The procedure was repeated with all possible combinations. Once when the problem ranking table is generated through the paired comparison priority ranking matrix table as shown in the results we further computed the RBQ as the percent ratio of the rank value by the sum frequency and VBI computed as usual and prioritised as the top most researchable problem where no solution exists for the problem present on date by definition.

The following steps were followed in administering the paired comparison technique:

1. Identify a set of problems, or preferences, to be prioritised

2. Note down each one of the problems on a separate card
3. Place two of these in front of the farmer and ask him to choose most preferred one with reasons for choice. Mark the response in the appropriate box in the priority ranking matrix
4. Follow the procedure for rest of the problems. Each time the criteria should be noted
5. Present a different pair and repeat the comparison
6. Repeat until all possible combinations have been considered
7. List the problems/preferences in the order
8. Repeat the pair wise ranking exercise for a number of individuals and tabulate their responses
9. The pair wise comparison table was prepared as below: The sum of frequencies of the problem under comparison scored over the comparing problem is taken as the basis to arrive at the final rank, where the order is represented by the highest sum of frequencies to the lowest in that order.

Results

An optimal combination of 66 stimuli generate 132 responses by the transposed matrix as compared to meeting 30 farmers ranking 12 problems which requires 360 stimuli and meeting the farmer individually, i.e. 30 times as compared to the paired comparison which can be administered to a group of 30 farmers at once, there is saving in the number of questions asked and responses which emerge redundantly costing us time and drudgery in administering the methodology for problem identification especially when participatory approach is followed, where the responses are supposed to be synergistic, cohesive and emerge on the basis of consensus and group think.

The results presented in Table 1 gives the number of problems listed by farmers on a consensus basis before it was subject to the paired comparison ranking procedure.

Table 1: Problems Ranking

1.	Fluorine in drinking water	-P1
2.	Non-availability of quality inputs in time	-P2
3.	Lack of quality seeds	-P3
4.	Poor water management	-P4
5.	Lack of support from devt dept	-P5
6.	Lack of fodder	-P6
7.	Lack of water for animals	-P7
8.	Diseases in cattle	-P8
9.	Flower drop in cotton	-P9
10.	Castor pod borer	-P10
11.	Redgram pod borer	-P11
12.	Water scarcity for crop production	-P12

The results presented in Table 2 give the problem priority ranking matrix completely computed and tabulated, the RBQ is computed as the percentage of sum frequency divided by the rank value which gives the RBQ and the VBI can be calculated in economic terms for a problem which varies temporally and spatially for similar problems expressed during the participatory field exercise for research problem identification and prioritisation.

Conclusion

Paired comparison technique of problem identification vis-à-vis the RBQ procedure hitherto followed by PRA practitioners in research problem identification not only saves time but eases the process and an alternative to arrive at the RBQ and VBI figures mentioned in the paper. It helps us to explore, understand people's decision-making process, needs, preferences (gender wise), ranking criteriae and priorities on a synergistic, holistic, participatory, cohesive, mode and emerge on the basis of consensus and group think.

It helps us to save time and the drudgery involved in problem prioritisation by the RBQ methodology wherein individual farmers rank the problems and frequencies and RBQ are computed later whereas it is done at once in the problem priority ranking matrix table computed in the paired comparison methodology. The agro-ecosystem analysis and Technology

Table 2: Problem Priority Ranking Matrix

Problem	P1	P2	P3	P4	P5	P6	P7	P8	P9	P10	P11	P12	T(f)	r(vlue)	RNK	RBQ
P1		P2	P3	P1	P5	P6	P7	P8	P9	P10	P11	P12	1	5	VIII	**20**
P2	P2		P3	P2	P2	P6	P7	P8	P9	P2	P11	P2	5	7	VI	**42.9**
P3	P3	P3		P3	P5	P6	P7	P8	P9	P3	P11	P12	4	6	VII	**33.3**
P4	P1	P2	P3		P5	P6	P7	P8	P9	P10	P11	P12	0	0	0	**0**
P5	P5	P2	P5	P5		P6	P7	P8	P9	P5	P5	P5	6	8	V	**75**
P6	P6	P6	P6	P6	P6		P6	P6	P9	P6	P6	P6	10	11	II	**90.9**
P7	P7	P7	P7	P7	P7	P6		P8	P9	P7	P11	P12	6	8	V	**75**
P8	P8	P8	P8	P8	P8	P6	P8		P9	P8	P11	P12	7	10	III	**80**
P9	P9	P9	P9	P9	P9	P9	P9	P9		P9	P9	P9	11	12	I	**91.7**
P10	P10	P2	P3	P10	P5	P6	P7	P8	P9		P11	P12	2	6	VII	**33.3**
P11	P11	P11	P11	P11	P5	P6	P11	P11	P9	P11		P11	8	9	IV	**77.8**
P12	P12	P2	P12	P12	P5	P6	P12	P12	P9	P12	P11		6	8	V	**75**
T(f)	1	5	4	0	6	10	6	7	11	2	8	6				
R(VLUE)	5	7	6	0	8	11	8	10	12	6	9	8				
RNK	VIII	VI	VII	0	V	II	V	III	I	VII	IV	V				
RBQ	**20**	**42.9**	33.3	**0**	**75**	**90.9**	**75**	**80**	**91.7**	**33.3**	**77.8**	**75**				

(Contd.)

Table 2 (Contd.)

Problem	P1	P2	P3	P4	P5	P6	P7	P8	P9	P10	P11	P12	T(f)	r(vlue)	RNK	RBQ
P1		P2	P3	P1	P5	P6	P7	P8	P9	P10	P11	P12	1	5	VIII	**20**
P2	P2		P3	P2	P2	P6	P7	P8	P9	P2	P11	P2	5	7	VI	**42.9**
P3	P3	P3		P3	P5	P6	P7	P8	P9	P3	P11	P12	4	6	VII	**33.3**
P4	P1	P2	P3		P5	P6	P7	P8	P9	P10	P11	P12	0	0	0	**0**
P5	P5	P2	P5	P5		P6	P7	P8	P9	P5	P5	P5	6	8	V	**75**
P6	P6	P6	P6	P6	P6		P6	P6	P9	P6	P6	P6	10	11	II	**90.9**
P7	P7	P7	P7	P7	P7	P6		P8	P9	P7	P11	P12	6	8	V	**75**
P8	P8	P8	P8	P8	P8	P6	P8		P9	P8	P11	P12	7	10	III	**80**
P9	P9	P9	P9	P9	P9	P9	P9	P9		P9	P9	P9	11	12	I	**91.7**
P10	P10	P2	P3	P10	P5	P6	P7	P8	P9		P11	P12	2	6	VII	**33.3**
P11	P11	P11	P11	P11	P5	P6	P11	P11	P9	P11		P11	8	9	IV	**77.8**
P12	P12	P2	P12	P12	P5	P6	P12	P12	P9	P12	P11		6	8	V	**75**
T(f)	1	5	4	0	6	10	6	7	11	2	8	6				
R(VLUE)	5	7	6	0	8	11	8	10	12	6	9	8				
RNK	VIII	VI	VII	0	V	II	V	III	I	VII	IV	V				
RBQ	**20**	**42.9**	**33.3**	**0**	**75**	**90.9**	**75**	**80**	**91.7**	**33.3**	**77.8**	**75**				

Action plan for the area can be quickly arrived at besides visually analysing the farmers' responses on the problem concerned when the paired comparison technique for problem identification is in progress.

Abbrvns used in Table 2:

T(f) = Total frequency of the problem expressed higher than the one being compared in pairs.

R Value = Rank value which is the reverse of the rank.

RBQ (obtained through paired comparison) = Rank Value (R)/T (f) expressed as a percentage.

Lessons Learnt from Watershed Management and its Development Programmes

15

N. Kalaichelvi and *E. Mahalakshmi Geetha*

Introduction

Water is the most precious natural resource and the universal asset. Water provides life supporting system for human beings, vegetation/plants and animals. Therefore, proper planning, development, management and optimal utilisation of water resources is of paramount importance for socio-ecological development. The twentieth century has seen phenomenal growth in the use of water. The world population has tripled, but the use of water for human purposes has multiplied six-fold. Today about 80 countries comprising 40 per cent of world's population suffer from serious water shortages.

Droughts are the results of serious imbalances in the hydrologic cycle leading to problem of water deficiency. The Ministry of Rural Development, with the help of NRSA, has identified 241 districts to be covered under Integrated Wasteland Development Programme (IWDP). The programme is presently being implemented by various state governments and NGOs in 146 districts. In the state of Tamil Nadu, number of districts is implementing DPAP and IWDP schemes. Soil and water conservation programmes play an important role in alleviating drought and rehabilitating wastelands.

Reducing per capita water availability

The average annual water availability in the country remains more or less fixed according to the natural hydrological cycle, but the per capita water availability is reducing exponentially due to increasing population. In India, the per capita water availability came down from 5300 m^3 in 1955 to 2200 m^3 in the early nineties against the world's average of 7400 m^3 and

Asian average of 3240 m^3. As per UN standards, the countries with annual per capita water availability of less than 1700 m^3 are considered as water stressed and those with less than 1000 m^3 as water scarce.

Where is the water shortage?

It is the fact that India is one of the most well endowed nations in the world in term of average annual rainfall; there is no reason why we should suffer from drought and shortage of water. It does not matter how much rain we get. It can be said that every village in India can meet its own water needs through proper management of natural resources like water. This can be achieved through water harvesting preferably by having small dams than mega structures. In simple words, 10 tiny dams with catchment of 1 ha each will collect much more water than one larger dam with catchments of 10 ha.

Why watershed management?

Loss of vegetative cover followed by soil degradation through various forms of erosion have resulted into lands, which are thirsty in terms of water as well as hungry in terms of soil nutrients. Watershed management (WSM) has, therefore, emerged as a new paradigm for planting, development and management of land, water and biomass resources with a focus on social and institutional aspects apart from biophysical aspects following a participatory "bottom-up" approach. Watershed management becomes increasingly important as a way to improve livelihood of people while conserving and regenerating their natural resources.

Concept of Watershed Approach

Over the last 50 years, there have been variations in conceptual models, objectives and implementation models of watershed programmes. Watershed management is a single window, integrated, participatory and sustainable area development programme. Watershed concept is an integrated approach of harmonised use of natural resources like land, water, vegetation, livestock, fisheries and human resources. Watershed, a hydrological unit of an area draining to a common outlet point, is recognised an ideal unit for planning and development of land, water and vegetation resources. Integrated watershed management covering the area from the highest point (Ridge line) to the outlet is, therefore, the process of formulating, implementing and managing a course of actions involving national and human resources in a watershed, taking account of all the operating within the watershed.

Watershed Programmes: Growth and Development

Multiple agencies have been involved in the growth and development of watershed programmes. Biophysical aspects of watershed management were started in the Damodar-Barakar basin under Damodar Valley Corporation, Hazaribagh by establishing a Soil Conservation Department during 1949-50. The centrally sponsored scheme of "Soil conservation in the catchments of river valley projects" was launched by the soil and water conservation division of the Union Agriculture Ministry in 1961-62 for watershed production in 27 catchments. The people's participation was the key to the success of these projects. During this period between 1980-81, watershed programmes were initiated under Flood Prone Rivers Project by the oil and water conservation division of Union Agriculture Ministry.

Based on the successful results of model watersheds in drought alleviation during the drought of 1987, number of nationally and internationally funded watershed development projects/programmes started. Major nationwide watershed programmes included DPAP, DDP that adopted watershed approach from 1987, NWDPRA in 1991 and IWDP on watershed basis since 1994. Apart from these national/international level programmes, a number of state government sponsored and NGOs supported watershed programmes subsequently started during 1980s and 1990s.

NWDPRA—National Watershed Development Project for Rainfed Areas

The Department of Agriculture and Cooperation, Ministry of Agriculture, Government of India is implementing a centrally sponsored scheme of NWDPRA in 25 states and two units of the country from the eighth plan onwards. The main objectives of the NWDPRA scheme are conservation, development and sustainable management of natural resources including their use, enhancement of the production and productivity of rainfed areas in a sustainable manner and restoration of ecological balance in the vast tracts of the rainfed areas. The restructured NWDPRA (WARASA Jan Sahbhagita guidelines) allows a much greater degree of flexibility in choice of technology, decentralisation of procedures, provision of sustainability and ensures active participation of watershed community in the planning and execution of their watershed development programme.

Model Watershed Programme—Experiences

Experiences gained

1. The judicious and capability based management of natural resources, namely soil, water, plant and animal on watershed basis is the key for maximising production on sustained basis by preservation of environment.
2. Watershed management should be a long-term policy for mitigating drought and flood and conservation of natural wealth.
3. Involvement of farmers, technicians, development agencies and administrators is essential at planning as well as at the development stage of watershed.
4. Water storage acts as a catalyst to win the hearts of farmers for the achievement and effective involvement of manifold increase in crop production and decrease in erosion hazards.
5. The approach of integrated management of resources is an utmost necessity, as no single component will produce long-term benefits.
6. The common developmental activities like community water storages, forestry, wells, roads, etc. require government support.
7. The necessary funds, inputs and technology may be made available at the proper time and place.

Lessons Learnt

The following constraints, feedback and research gaps have been identified from the model watershed experiences.

Constraints

1. Lack of investment in dryland areas appears to be a major constraint in development of drylands on a permanent basis.
2. The poverty of farmers and lengthy learning procedure create problems in the implementation of the programme.
3. The flow of funds from Government of India-State-District line department is sometimes not smooth.
4. Sometimes there are poor linkages and cooperation among the organisations involved.
5. Lack of trained and experienced staff with the departmental agencies. This is one among the key constraints.
6. Difficulties in timely supply of inputs, namely seed, fertilizers, etc.

Feedback

1. All the developmental activities namely, engineering works, forestry measures, agronomic practices, etc. must be taken up with equal emphasis.
2. Biological measures may be preferred over conventional engineering measures in agricultural lands with slight to mild slopes.
3. The duration of the project for perfect implementation and evaluation should not be less than 7 to 10 years considering the slow flow of benefits from trees and slow acceptance of technology by the farmers.
4. Provision of funds may be made for maintenance and care of soil conservation works, etc.
5. Revision and improvement in master plans may be made under basis of experience gained while working in the watersheds.

REFERENCES

Shri Madhu M., Central Soil and Water Conservation Research and Training Institute, Research Centre, Udhagamandalam.

Shri Sikka, A.K. and Samra, J.S. (2000). "Participatory Watershed Management—A New Paradigm for Integrated Water Resource Management" Proc. Inter. Conference on Integrated Water Resource Management for Sustainable Development, New Delhi.

16

Capacity Building Measures in Management of a Successful Watershed in Coimbatore District

K. Palanisami, A. Raviraj and *S. Gurunathan*

Introduction

Groundwater is the main source for rural water supplies in many developing countries. Over recent years, increasing abstraction to meet rising demand for domestic supplies and irrigation has raised concerns for the sustainability of the resource and the livelihoods it supports. Consequences of over-exploitation include declining water levels and increasing competition for scarce water resources between domestic and agricultural users and rural and urban communities.

Considerable emphasis is being given to the augmentation of natural recharge by both traditional and modern techniques. Some of these techniques have been employed for centuries, ranging from simple check bunds in gullies to complex diversion and infiltration structures as well as injection wells. Recently there has been considerable investment and renewed effort to restore and maintain such traditional facilities as well as building new structures. Much of the current effort is, however, empirical in choice of sites, structures and aquifers.

Kodangipalayam watershed in Palladam block of Coimbatore district was taken for the study of artificial recharge to understand the natural and artificial recharge processes occurring in the main recharge structure and the satellite recharge structures.

Community Organisation and Entry Point Activities in the Watershed

Participatory Rural Appraisal (PRA) is one of the most important exercises in watershed development projects before finalisation of Action Plan. In fact, Kodangipalayam watershed was selected after due PRA and the watershed with greater participatory response would be preferred. After sanction of the project, the WDT toured extensively and organised various UGs and other groups. As a part of confidence building exercise, some community benefiting entry points activities were taken up by the PIA directly. The treatment plan and the interventions were decided after elaborate PRA exercise. These activities were drinking water sources/wells, hand pumps, bathing ghats, approaches to water tanks, village roads, village sanitation improvement works, etc. Water harvesting measures and improvement of drinking water sources were given more importance.

Maintenance and Management of the Structure

Group formation

The construction of AR structure and the watershed development activities created awareness among farmers to formulate various groups and associations to perform collective action, to ensure people's participation for various watershed development activities like construction of ponds and structures, water harvesting structures, for lobbying with government machinery to get maximum possible benefits to their watershed. A large number of user groups were formed. The details are given in Table 1.

The Project Implementing Agency (PIA) constituted User Groups (UGs) in the watershed areas with the help of Watershed Development Team (WDT). These groups are homogenous groups, who may be most affected by each work/activity and shall include the persons having landholding within the watershed areas. Each UG shall consist of the persons who are likely to derive direct benefits from a particular watershed work or activity. The UGs should actually takeover the operation and maintenance of the completed community works or activities on common property resources.

Capacity building and training

The DRDA and the PIA organised relevant training programmes for all the functionaries involved in watershed development and a majority of the members of the UGs were given basic training involving skill up gradation and orientation on the technical and organisational aspects of the running of these groups. The UG members were taken for visits to research stations/successful watersheds/Kisan Vikas Kendras, etc. where they could see the demonstration of successful technologies/ practices/designs that are relevant to them. The results of the study conducted about the awareness developed among the farmers in the watershed villages about various watershed development activities and various treatment activities that were performed under the Kodangipalayam DPAP watershed project during 1995-99. The results are presented in Table 2. The response was very high that about 70-90 per cent of the respondents were aware about various treatment activities done in the project.

Collective action

The degree of collective action performed by the respondents is shown in Table 3. The respondents were average to good in performing various developmental activities collectively during the phase of treatment activities. But when it was about the post-maintenance of structures already built they were lacking in commitments, co-operation and collective action. At this stage they should be educated about the need for collective action to maintain the structures built in their own locations which were built for their own benefit.

Adoption of technologies

It is important and interesting to note that about 100 of the respondents accepted the importance of water harvesting structures in their households and their fields (Table 4). But when we come to adoption of micro irrigation structures as the respondents are resource poor farmers they were unable to afford the cost of installation of those costly structures and as the scope for irrigated agriculture in the study area is getting diminished nowadays the farmers were not ready to install these micro irrigation structures.

Water conservation methods

With the mounting necessity for and scarce availability of groundwater, there is an awful prerequisite for development of water harvesting structures. These factors oblige the farm households to construct water harvesting structures like farm ponds, livestock ponds, check dams, planting trees, etc. in their fields. One of the crucial structures found in the study area is structure of farm surface storage tanks. Uneven and erratic power supply, very low water table urge the farmers (particularly bore-well farmers owning compressor type motors) to build up farm surface storage tanks. The water is pumped from very rumbling bore-well and stored in these tanks and the stored water is used for irrigating crops which is presented in Table 5.

It was evident that more water conservation methods were adapted in with area (27.50 per cent) than without area (19.75 per cent). The farmers in the recharge structure area realised the significance of water conservation methods.

Impact assessment analysis—respondents' opinion

The overall impact of watershed treatment activities and the artificial recharge structure constructed in the village stated by the farm households is presented in Table 6. Of the different beneficial impacts, the watershed development activities and the construction of artificial recharge structures exert more impact on groundwater recharge, increase in water table and increase in irrigated area.

Table 1: Group Formation Activities in the Watershed Area

Sl. No.	Particulars	Before	After
1.	NGOs	—	1
2.	User groups	—	14
3.	Groups recovering revolving funds properly	—	3
4.	WUA	—	4

Table 2: Awareness/Usefulness of the Programme (Percentage of the Respondents)

Sl. No.	Programmes	Aware	Not aware
1.	Watershed activities	90	10
2.	Summer ploughing	70	30
3.	Contour bunding	65	35
4.	Farm ponds	85	15
5.	Percolation ponds	72	28
6.	Check dams	67	33
7.	Activities of DRDA/DPAP	60	40

Table 3: Degree of Extent of Collective Action

Sl. No.	Activities	Poor	Average	Good	Very good
1.	Construction of recharge structures			√	
2.	Post-maintenance of structures	√			
3.	Repair works		√		
4.	Maintenance of natural resources		√		
5.	Desilting of tanks			√	

Table 4: Adoption of Technologies in Kodangipalayam Watershed (Percent)

Sl. No.	Technologies	Adopting	Not adopting
1.	Rainwater harvesting	100	—
2.	Summer ploughing	65	35
3.	Water conservation methods (micro irrigation, etc.,)	10	90
4.	Altering cropping pattern	100	—

Table 5: Water Conservation Methods Adopted by Sample Farmers in the Study Area (percent)

Sl. No.	Particulars	With	Without
1.	Surface Storage tank	5.00	12.50
2.	Farm Pond	5.00	—
3.	Modern Irrigation techniques	12.50	5.00
4.	Others	5.00	2.25

Table 6: Respondents' Opinion about Overall Impact of Watershed Development Activities and the Recharge Structure (Percentage of Respondents)

Sl. No.	Impacts	Yes	No	Not sure
1.	Increase in recharge	100	—	—
2.	Increase in water table	100	—	—
3.	Increase in irrigated area	60	25	15
4.	Changes in cropping pattern	65	22	3
5.	Increase in yield	45	15	40
6.	Improve soil fertility	38	38	24
7.	Improve water quality	55	30	15
8.	Increase in green cover	60	40	—

Role of Project Implementing Agency (PIA) in Watershed Development Programme

17

P. Balaji and *S. Gurunathan*

Introduction

The objective of any development programme or project is to transform a set of resources (inputs) into desired results or outcomes (outputs). Understanding the nature, objectives and scope of the development project and the responsiveness of target groups is an imperative for development personnel/specialist, economists and policymakers to recommend improvements that will guarantee more food, fodder, fuel and livelihood security for resource poor farmers who form the majority of the farming community and the landless poor.

Implementing agency would see a greater value in spending an extra few crores of rupees for undertaking works in the field rather than spending this money for monitoring and evaluation. However, according to some observers, mid-course corrections can improve the programme benefits substantially, in some cases up to 100 per cent. But even if we consider the improvement to be very modest, say, 10 per cent, then a 1 per cent of programme outlay on meaningful monitoring and evaluation would have a very high pay-off in terms of achieving the programme objectives. It is of utmost importance, therefore, to put in place an institutional mechanism for research and monitoring and evaluation in the field of watershed development by involving reputed institutions in the country for upgrading the quality of evaluation. For example, evolving methodologies for capturing the participatory processes and convergence of complementary interventions, by guaranteeing to them adequate funding on a long-term basis and by ensuring for them autonomy in the pursuit of monitoring and evaluation.

Systematic feedback of information was collected from the project areas and beneficiaries for whom the project is intended. To provide the project management with such information, it is essential to understand the results of their activities for which data should be gathered continuously and analysed without delay. A system combining such data gathering, analysis and interpretation is called a monitoring and evaluation system.

Objectives

(i) To study the role of PIA

(ii) The problems faced by farmers and watershed association and

(iii) To analyze the Feedback

The Project Implementing Agency (PIA) plays an important role in the implementation of the various activities proposed under the watershed development programmes. The members of the User Groups are expected to have frequent interaction and to work in collaboration with each other. Such an interaction fully depends upon how the members of User Groups perceive the role played by PIA. Therefore, it was pertinent to study the perception of the user groups on PIA. The opinion on the technical support extended by the PIA is given in Table 1. It is lucid from the table that nearly 80 per cent of the respondents found the technical support extended by the PIA was good, 16 per cent as average and negligible (2%) found it as poor. Watershed-wise details indicate that the technical support extended by PIA was more in Kovilur and Kanuvaipudur.

The opinion of the respondents about PIA in relation to the accessibility of PIA was analysed. Of the total respondents, around 60 per cent found that there was good access to PIA officials, 32 per cent found average and only 6 per cent found poor. Of the different watersheds studied, accessibility to PIA was good in Ernapuram registering a highest percent of 73 per cent (Table 2).

The most important problem prevents the farmers in adopting technologies is the small size of holding as the mean score is highest in all the watersheds. The other important problems are labour problem and water scarcity (Table 3). It is evidenced from the table that the small size of holding is more in the case of Kanuvaipudur watershed. The other factors identified viz., lack of technical support and follow-up action and irrelevant technologies form the least influencing problems which hinder adopting technologies.

Enough efforts have also been made to study the problems faced by the watershed association president and the officials of the watershed committee. As these officials are playing major role in implementing various watershed development activities, the study of their problems are very important. Of the different problems studied, inadequate funds form the major problems faced by these officials. This is followed by lack of villagers co-ordination and co-operation and inadequate time for planning and implementation of various watershed development activities (Table 4).

Out of the respondents 82.67 per cent have attended the training, while 17.33 per cent of them have not attended. Watershed-wise details show that in almost all the watersheds about 80 per cent of the respondents have attended the training except in Ernapuram (66.67 per cent) and Thangayur (76.67 per cent). Awareness should be created among the rest of the respondents, to attend these training programmes, as it will motivate them to participate in watershed development in a positive way (Table 5).

It is evidenced that 80 per cent of the respondents found that the training programme was useful for them. Watershed-wise details show that majority of the farmers felt that the training programme was useful. In these watersheds, insignificantly described and found that the training not so useful in carrying on the various watershed development activities and technologies introduced (Table 6).

They felt so because the duration of the training conducted was insufficient and the training materials supplied and discussed during training were also not adequate. Though the majority of the beneficiaries are satisfied with the training programme, a small fraction of beneficiaries expressed discontent over the training programmes. This suggests that the training programme needs to be tailored to meet the requirements of beneficiaries.

Opinion of the respondents with regard to the functioning of User Groups (UGs) was more than 80 per cent of the respondents expressed that the UGs functioned good and nearly 12 per cent were found that average. Among the different watersheds the UGs are functioning good in Kovilur watershed (Table 7).

Among the total respondents, nearly 88 per cent reported that the participation was good, 10 per cent average and only meagre (1.33 per cent) reported poor. Though majority of respondents participated in general meetings is appreciable, it is suggested that enough steps have to be taken to involve all the members in general meeting, especially in those watersheds where the members' participation in the User Group meeting was found to be weak and average (Table 8).

The interaction of members of User Groups with Watershed Association/Watershed Committee is a must for the success of watershed development programme. Nearly 89 per cent of the respondents reported that there was a good interaction of members with the watershed association/watershed committee (Table 9). Only very few reported that average (6 per cent) and poor (4.67 per cent).

The interactions of respondents with officials of Project Implementing Agency indicate that there was good interaction. It is evidenced that nearly 67 per cent of the respondents found that good interaction, 22 per cent as average and 10 per cent with poor (Table 10).

Conclusion

The progress of the project implementation has been monitored through periodical, quarterly reports. Monthly review meetings are conducted by the Project Officer and the Collector and Chairman, DRDA and by the Project Implementing Agency. The quality of execution works are monitored by the Community. Frequent inspections and visits are conducted by officials at various levels. An impact evaluation study was conducted in June 1999 by COODU, an experienced NGO.

This agency has also made a video film 'Pasumai Thakkam' for DRDA Salem covering the project activities and benefits realised by the watershed communities.

Table 1: Respondents Opinion about PIA in Provision of Technical Support

Watersheds	Good	Average	Poor	Total
Chinnasoragai	23 (76.67)	7 (23.33)	0 (0.00)	30 (100.00)
Ernapuram	25 (83.33)	4 (13.33)	1 (3.33)	30 (100.00)
Kanuvaipudur	26 (86.67)	4 (13.33)	0 (0.00)	30 (100.00)
Kovilur	27 (90.00)	3 (10.00)	0 (0.00)	30 (100.00)
Thangayur	20 (66.67)	7 (23.33)	3 (10.00)	30 (100.00)
Total	121 (80.67)	25 (16.67)	4 (2.67)	150 (100.00)

Table 2: Respondents' Opinion about Accessibility to PIA

Watersheds	Good	Average	Poor	Total
Chinnasoragai	15 (50.00)	15 (50.00)	0 (0.00)	30 (100.00)
Ernapuram	22 (73.33)	6 (20.00)	2 (6.67)	30 (100.00)
Kanuvaipudur	19 (63.33)	8 (26.67)	3 (10.00)	30 (100.00)
Kovilur	20 (66.67)	10 (33.33)	0 (0.00)	30 (100.00)
Thangayur	15 (50.00)	10 (33.33)	5 (16.67)	30 (100.00)
Total	91 (60.67)	49 (32.67)	10 (6.67)	150 (100.00)

Table 3: Problems Faced by Farmers in Adopting Technologies

Particulars	Chinna-soragai	Ernapuram	Kanuvai-pudur	Kovilur	Thangayur
Small size of holding	35.3	44.1	43.6	42.3	35.5
Water scarcity	9.8	14.5	12.7	12.0	10.8
Labour problems	12.8	15.9	19.8	15.9	15.2
Lack of technical support and follow-up action	0.8	1.0	0.8	0.7	0.8
Irrelevent technologies	0.7	0.8	0.9	0.6	0.9

Note: Figures indicate worked out mean score employing Garrett ranking technique.

Table 4: Problems Faced by the Watershed Association President/Secretary

Particulars	Chinna-soragai	Ernapuram	Kanuvai-pudur	Kovilur	Thangayur
Inadequate time for planning and implementation	8.5	7.6	5.8	8.6	5.9
Lack of prior information from the project staff	0.8	0.8	0.3	0.5	0.1
Lack of villagers co-ordination and co-operation	19.6	17.3	18.7	16.0	15.8
Inadequate funds	56.8	65.2	62.8	58.7	49.8
Political interference	0.9	1.2	0.7	1.3	1.1

Note: Figures indicate worked out mean score employing Garrett ranking technique.

Table 5: Training Programme Attended by the Respondents

Watersheds	Attended	Not Attended	Total
Chinnasoragai	25 (83.33)	5 (16.67)	30 (100.00)
Ernapuram	20 (66.67)	10 (33.33)	30 (100.00)
Kanuvaipudur	27 (90.00)	3 (10.00)	30 (100.00)
Kovilur	29 (96.67)	1 (3.33)	30 (100.00)
Thangayur	23 (76.67)	7 (23.33)	30 (100.00)
Total	124 (82.67)	26 (17.33)	150 (100.00)

Table 6: Usefulness of the Training Programme as Perceived by the Respondents

Watersheds	Useful	Not Useful	Total
Chinnasoragai	23 (76.67)	7 (23.33)	30 (100.00)
Ernapuram	20 (66.67)	10 (33.33)	30 (100.00)
Kanuvaipudur	26 (86.67)	4 (13.33)	30 (100.00)
Kovilur	29 (96.67)	1 (3.33)	30 (100.00)
Thangayur	22 (73.33)	8 (26.67)	30 (100.00)
Total	120 (80.00)	30 (20.00)	150 (100.00)

Table 7: Opinion of Respondents about the Functioning of the User Groups

Watersheds	Good	Average	Poor	Total
Chinnasoragai	25 (83.33)	5 (16.67)	0 (0.00)	30 (100.00)
Ernapuram	26 (86.67)	4 (13.33)	0 (0.00)	30 (100.00)
Kanuvaipudur	27 (90.00)	3 (10.00)	0 (0.00)	30 (100.00)
Kovilur	30 (100.00)	0 (0.00)	0 (0.00)	30 (100.00)
Thangayur	23 (76.67)	7 (23.33)	0 (0.00)	30 (100.00)
Total	131 (87.33)	19 (12.67)	0 (0.00)	150 (100.00)

Table 8: Opinion of the Respondents about the Participation in User Group Meetings

Watersheds	Good	Average	Poor	Total
Chinnasoragai	27 (90.00)	3 (10.00)	0 (0.00)	30 (100.00)
Ernapuram	26 (86.67)	3 (10.00)	1 (3.33)	30 (100.00)
Kanuvaipudur	28 (93.33)	2 (6.67)	0 (0.00)	30 (100.00)
Kovilur	27 (90.00)	3 (10.00)	0 (0.00)	30 (100.00)
Thangayur	25 (83.33)	4 (13.33)	1 (3.33)	30 (100.00)
Total	133 (88.67)	15 (10.00)	2 (1.33)	150 (100.00)

Table 9: Interaction with Watershed Association

Watersheds	Good	Average	Poor	Total
Chinnasoragai	28 (93.33)	2 (6.67)	0 (0.00)	30 (100.00)
Ernapuram	29 (96.67)	1 (3.33)	0 (0.00)	30 (100.00)
Kanuvaipudur	27 (90.00)	3 (10.00)	0 (0.00)	30 (100.00)
Kovilur	30 (100.00)	0 (0.00)	0 (0.00)	30 (100.00)
Thangayur	20 (66.67)	3 (10.00)	7 (23.33)	30 (100.00)
Total	134 (89.33)	9 (6.00)	7 (4.67)	150 (100.00)

Table 10: Interaction of Respondents with PIA officials

Watersheds	Good	Average	Poor	Total
Chinnasoragai	19 (63.33)	7 (23.33)	4 (13.33)	30 (100.00)
Ernapuram	26 (86.67)	4 (13.33)	0 (0.00)	30 (100.00)
Kanuvaipudur	24 (80.00)	5 (16.67)	1 (3.33)	30 (100.00)
Kovilur	23 (76.67)	7 (23.33)	0 (0.00)	30 (100.00)
Thangayur	9 (30.00)	10 (33.33)	11 (36.67)	30 (100.00)
Total	101 (67.33)	33 (22.00)	16 (10.67)	150 (100.00)

Sustainability of Watershed Development

18

M. Malini and *M. Ranichandra*

Watershed

The word watershed refers to a "contiguous area draining into a single water body or a water course" or "it is a topographical area having a common drainage". This means that the rainwater falling on an area coming within a ridgeline can be harvested and will flow out of this area through single point. Some refer it as a catchment area or river basin. A watershed can be defined at various levels, at a macro-level (district or regional levels), at village level (hundreds or thousands of hectares), at farm level or even within the farm.

Watershed Development in India

Approximately 170 million hectares in India are classified as degraded land, roughly half of which falls in undulating semi-arid areas where rainfed farming is practised. Growing efforts—by the state and NGOs—to reverse this decline are often based on the rehabilitation of micro-watersheds. The micro-watershed concept aims to 'establish an enabling environment for the integrated use, regulation and treatment of water and land resources of a watershed based eco-system to accomplish resource conservation and biomass production objectives (Jensen et al., 1996).

Although a micro-watershed may be a sensible planning unit from a biophysical perspective, yet many would argue strongly against the appropriateness of such a unit for rural development. Social institutions to promote cooperation—necessary for the protection and rehabilitation of private and common pool resources—are usually village-based. Often however biophysical and socio-political boundaries do not coincide. The majority of projects do not, therefore, strictly entail the development of "a watershed", rather they adopt an approach to rural development incorporating principles of the watershed approach.

Opinion of Watershed Expert

In the words of eminent economist, C.H. Hanumantha Rao, "Watershed development has been conceived basically as a strategy for protecting the livelihoods of the people inhabiting the fragile ecosystems experiencing soil erosion and moisture stress."

Sustainability of Watershed Development

The objectives of watershed programmes combine a range of targets which include:

- *The productivity objective*: The generation of benefits (either through better management of existing resources and prevention of further degradation and/or through schemes to invest in improvement of the resource.
- *The equity objective*: A fairer distribution of benefits from the resource and their spread to a wider group of people.
- *Sustainability objective*: Institutional and ecologically sustainable development at local and national levels.

Productivity

Given their relatively recent history and the challenges which watershed development programmes have set themselves, it is difficult to make a detailed assessment of the performance of watershed development programmes. It is perhaps not surprising that there have been few economic appraisals of completed projects due to their complex nature (Kerr et al., 1996b). A useful compilation of studies is found in the July-September 1991 issue of the Indian Journal of Agricultural Economics. They estimate overall returns to investments in watershed development to be generally high, with cost-benefit ratios ranging from one to greater than two.

Sharing the benefits—Issues of equity

Evidence suggests that certain social groups have been consistently marginalised by community development initiatives

(Hinchcliffe et al., 1995; Hobley, 1992). Studies suggest that this is also true of watershed development projects. In a case study of MYRADA's project in Karnataka for example, Fernandez (1993) identifies four groups in particular who do not seem to benefit from watershed development; the landless, families in the upper levels of catchments, women and marginalised tribal groups. Concerns over equity are perhaps most apparent where development efforts focus on the rehabilitation of common pool resources.

Drawing on experience with community involvement in forest management in Nepal and India, Hobley and Shah (1996) observe that inequalities tend to be most acute at the front and back end of the process of building common pool resource institutions. During the early stages, when CPR regimes are first introduced, the poor are affected most adversely. Their greater dependence on what were previously *de facto* open access resources, means that constraining access during the necessary period of environmental recuperation has a disproportionate impact on the poor. At the other end of the process, as the CPR regime matures, the increased value of the resource frequently attracts local (and not so local) commercial and political interests which also rarely benefit the poor.

Where watershed development has the explicit objective of providing greater access for poorer groups, such a shift in power is bound to be contested by those who lose out. The sustainability of such efforts are, therefore, intricately linked to changes in local institutional and power structures. The impact of watershed development efforts on women is also a key issue (see papers by Lokur Pangare and D'Souza in this mailing). Turton et al. (1997) note that access restrictions imposed on common grazing areas encourage a shift to stall feeding systems. The main bulk of the work of collecting fodder for livestock is undertaken by women, who have to spend extra time cutting and carrying feeding materials.

Allsebrook (1996) notes that "scaling up...tends to distance organisations further from the weakest members of society". It is, therefore, essential that agencies act to protect the interests of the poor and women. Although explicit efforts are often made to overcome their exclusion through community anticipatory techniques, yet the limited public airing of conflict which results is generally insufficient for achieving social consensus (Mosse, 1996). Mechanisms need to be identified which promote the interests of the disadvantaged and increase their access to resources.

Sustainability

Watershed development is fundamentally about the creation of new opportunities—both in an institutional sense (through an increased capacity of local communities to organise and protect their resource base and to draw on the services of external organisations) and ecologically (in terms of increases in resource productivity and nutrient flows within the watershed).

The ultimate indicator of success is the ability of communities to take advantage of new opportunities and the extent to which benefits are sustained in the long-term. Communities need to be able to call on adequate technical help and advice from elsewhere, at least until such a time as they feel confident in their abilities to manage the improvements they have adopted.

Agricultural support programmes need to be strengthened and reoriented in order to maximise the benefits from other physical investment. As the benefits of SWC begin to show, the production possibility frontier is pushed outwards creating demand for new varieties, soil management technologies, etc. which can and should serve to stimulate long-term engagement with external agencies.

In the IGWDP, farmers are expected to obtain loans or invest their own funds in downstream improvements such as dairy production, horticulture, wells and new crop varieties to take full advantage of the SWC measures (Farrington and Lobo, 1997). Even where groups for joint action have been formed successfully over a period of years, they remain vulnerable to external 'shocks'. These include the emergence of political conflict within the group as members ally themselves differently to political parties in local elections or conflicts with other villages over management of the shared resources. To speculate that one-third of groups formed with NGO support collapse within five years of NGO withdrawal may not be unreasonable.

It is becoming increasingly apparent that links between watershed institutions and local political structures are essential for sustaining the new institutions. There have been numerous deliberate attempts by political bodies to undermine watershed development activities (Mukherjee, pers. comm.).

Watershed associations have become a hunting ground for political parties, partly as a result of the considerable funds at their disposal (Krishna, 1996). Sida (1997) noted the potential conflict arising as new leaders emerge in the villages and existing systems are challenged. In some villages, the watershed organisations are trying to influence the *panchayats* and other establishments that support development activities, pressuring them to explain the differences in wage rates and asking for more transparency in their way of working.

They conclude that the distinction between the long-term roles of the watershed organisation and the village *panchayat* warrant serious consideration by the project. The assumption is that the village organisations promoted for planning and implementation of watershed activities will be capable of maintaining both on-going and new activities into the future. However, NGOs—some with more than a decade of experience of watershed development through village organisations—have found that withdrawal after a fixed period can derail the process. Because of the importance of sustainability, planning for the graduation of the village, from a large degree of external dependence at the beginning of the project to a small amount at the end of four years, should start at the selection and planning stage.

This should include a strategy to bring in the external agencies, government and private, whose increasing involvement will be needed as the project progresses. This process would enable the villagers to learn about these agencies, about what they have to offer in the way of help and advice and how to utilise that help and advice.

Conclusion

The success of participatory watershed development will be judged according to the extent to which improvements in productivity are both equitable and sustainable. The key question is whether the scaling up of participatory approaches, necessitating a move away from resource intensive approaches to the formation of new partnerships, undermines the extent to which this is achieved.

REFERENCE

Cathryn Turton, Michael Warner and Ben Groom, (1998). Scaling up Participatory Watershed Development in India: A Review of the Literature, *Agricultural Research and Extension Network*, Paper no. 86.

19

Motivation and Adoption of Agroforestry (*Wadi*) in the Peri-Urban Village Channapur

Anasuya Patil, G.H. Yogesh and *C.S. Hunshal*

Introduction

Agriculture forms an important sector in the Indian economy. About 70 to 75 per cent of the working population depends on agriculture for their livelihoods. Pace of adoption of an innovation and consequent diffusion on large scale is the essential feature of development. Adoption of innovations depends on personal, psychological, social and economic aspects of farmers play a major role in their adoption decisions. Motivation is of social origin that arises out of human relations within and between work groups of informal structure. The agriculture, which was once, self-sufficient for its input requirements, became dependent on more and more external, costly and scarce inputs. With advanced thinking on modern economic lines and increased use of modern scientific technologies, a school of thought emerged which wanted agriculture to be an industry.

The project entitled "Enhancing Livelihoods and Natural Resource Management in Peri-Urban Villages" made a beginning in project area to establish *wadi* on a participatory planning and implementation and evaluation in order to enhance the natural resource management. In the participatory process the farmer's needs were assessed, based on the needs they were trained, taken on exposure visits and focused on participatory approach for implementation of *wadi* (integration of many components of farming especially the tree component). The establishment of *wadi* was appropriate for the PUI research because of only 185 landholding families of which 153 were small and marginal landholders and were growing only agricultural crops. Farmers felt that establishment of *wadi* would be able to sustain the agriculture for many reasons viz., the trees have been a source of wood, food, fuel, fibre, firewood, fodder, manure and other minor products and also prevents soil erosion. Recent years the concept of growing trees on agricultural land known as agroforestry has received much attention due to neglect of tree component in agriculture owing to monocropping.

Keeping the above facts in view, the present investigation was designed with the following objectives:

1. To study the motivational pattern of farmers.
2. To measure the knowledge level of farmers before and after as a result of *wadi* intervention.
3. To find out relationship between knowledge and adoption behaviour of *wadi* farmers.
4. To determine the constraints faced by the non-*wadi* adopters.

Materials and Methods

The study was carried out in Channapur village of Hubli taluk, Dharwad district in Karnataka state in South-West of India during the year 2004 with a sample of 47 farmers. All the farmers who were involved in participatory planning on *wadi* were purposively selected and data was collected with the help of pre-structured schedule, which was designed separately for *wadi* adoptors and non-adoptors. The family information survey and participatory wealth ranking was collected before the intervention. The motivation pattern was collected through the schedule and respondents were categorised into various motives and ranks were assigned. Focus group discussions were carried out involving both men and women to understand the perceived constraints faced by the non-adopters.

The data was collected before and after the intervention by personal interview method in an informal atmosphere. The extent of knowledge before and after the intervention was studied with the help of teacher-made test developed for the purpose. The respondents were categorised into low and high knowledge level by taking mean knowledge index as a measure of check. Depending upon the adoption scores the farmers were grouped into high and low by using mean as measure of check for adopters and non-adopters. Data was processed by using knowledge index and adoption quotient. Data was analysed by both qualitative and quantitative types and coded, tabulated using excel, correlation coefficients, Z and paired t test.

Results and Discussion

1. *Motivation pattern of farmers*

It could be seen from the Table 1 that highest score (189) of the respondents indicated the utilitarian motive as the motive that motivated them to adopt *wadi*. The other motives felt by the large number of respondents to adopt *wadi* include economic motive (178 score) and it was followed by achievement motive (147 score), affiliation motive (110 score) and innovative motive (81 score).

Table 1: Motivation Pattern of Farmers

(n=47)

Motives	Motivation Score	Rank
Utilitarian motive	189	I
Economic motive	178	II
Achievement motive	147	III
Affiliation motive	110	IV
Innovative motive	81	V

However, it was interesting to note that the innovation motive was felt as least important in adoption of *wadi*. With respect to adoption of *wadi*, it could be seen that most of the *wadi* farmers stated utilitarian motive as most important motive that inspired them to adopt *wadi*. The availability of vegetables, fruits for home consumption and resources for future generations might have been a strong influencing factor that motivated the farmers to adopt *wadi*. The next in the order of motivations expressed by farmers happens to be economic motive. *Wadi* is like a cash crop having fruits that have more demand in the urban market. The necessary demand from urban areas, extrinsic motive and regular returns with lesser investments might have motivated the farmer to adopt *wadi*. Achievement motive was preferred as the third important motive, because the farmers wanted to utilise their available resources including the by-products from farm, vermicompost and efficient use of labour and land.

The affiliation motive was also stated by considerable number of farmers as this motive influenced them to adopt *wadi* and was ranked fourth in the order of preference. The reason might be that the affiliation is the social capital and that of individuals is to face new experiences and bear new risks by adopting *wadi*.

Assigning lower ranks to the innovative motive as a motive to adopt *wadi* might be due to the lack of urge to face new experiences. On the basis of psychological needs it has been observed that there are various types of motives such as utilitarian motive, which plays a significant role in the acceptance of new technology. In most of these cases the cognitive tension or search for meaning or personal significance drives one to accept an innovation. Thus, it may be stated that, the study of motivation takes for granted that disposition to behave in certain ways could be changed by practice.

2. *Knowledge index of farmers as a result of intervention*

The content of Table 2 reveals that the net gain in knowledge was 54.17 for low category and 67.83 for high category. It was subjected to "paired t" test and was found significant at 1 per cent level.

Table 2: Knowledge Index of Farmers as a Result of Intervention

(n=47)

Category index	Knowledge index				Total value	t value
	No. of respondents	Before inter-vention	After inter-vention	Net gain		
Low	22	23.50	47.00	77.65	54.17	69.85**
High	25	18.70	53.00	86.50	67.83	47.99**
Overall	47	41.00	100.00	82.40	61.40	47.21**

** Significant at 1% level.

The reasons could be that the higher consultancy pattern of BAIF and UAS research partners and also greater participation of farmers in the activities such as meetings, trainings, exposure visits, discussions, field trips and demonstrations. Consequent to their more contacts with BAIF staff (placed in the village) and UAS scientists (once in a week) and higher participation in various participatory methods might have accumulated great deal of knowledge on *wadi*. These findings are supported by the studies conducted Dharminder and Ravinder (2004); study revealed that early maturing varieties were known by 90 per cent of trained and 61.3 per cent of non-trained cotton growers. Majority of trained cotton growers reported recommended dosage of nitrogen and phosphorus. Further majority of trained cotton growers had a good knowledge of natural enemies. Van de Fliert (1998) also reported that FFS training generally resulted in increased knowledge level of farmers and they become "Ecosystem managers" instead of "Package adopters."

3. *Pattern of adoption of wadi practices followed by farmers*

Practice-wise pattern of adoption of *wadi* was studied and presented in Table 3. All the farmers adopted the method of filling pit with FYM and planted horticulture plants such as mango and sapota. Whereas majority of them (96.96%) followed digging pits of size 3 × 3 × 3 ft followed by filling pit (90.90%) with green leaves and planting forest species as timber-fodder-fuelwood and stylosanthus as a leguminous fodder source on contour bunds and application of tank silt (87.87%). The trend was not encouraging with respect to farmers' indigenous practices (6.06%), vermi-compost (33.33), growing of vegetables on bunds around farm ponds (51.51%) and digging farm ponds (57.57%). It was encouraging to note that farmers followed *wadi* practices to a considerable extent by mulching ring around the plant with farm waste, i.e. agriculture by-products (81.81%), soil and water conservation (81.81%), drip irrigation by earthen pot method (78.78%), ring-cum-trench method by filling trench with farm waste (78.78%).

Wadi adopters expressed their happiness over the benefits they could accrue from the agroforestry. Farmer's preferred mango trees because of better market and higher price compared to sapota and also the higher survival rate. Similarly, under the forestry plants the highest rate of survival has been in case of

Table 3: Pattern of Adoption of *Wadi* Practices by Farmers

(N=33)

Adoption of *Wadi* Practices	No. of Farmers Followed the Practices	
	Σf	%
Digging of pits by 3 × 3 × 3 ft	32	96.96
Filling pit with green leaves	30	90.90
Filling pit with vermi-compost	11	33.33
Filling pit with FYM	33	100.00
Tank silt	29	87.87
Horticulture: Mango-Sapota	33	100.00
Forest Species: Timber-Fodder-Fuelwood	30	90.90
Drip irrigation by earthen pot method	26	78.78
Ring-cum-Trench method by filling trench with farm waste	26	78.78
Mulching of ring around the plant with farm waste	27	81.81
Soil and water conservation	27	81.81
Farm ponds	19	57.57
Stylosanthus as a fodder source on contour bunds	30	90.90
Growing of vegetables on bunds around farm ponds	17	51.51
Farmers' indigenous practices	2	06.06

teak and silver oak. Stylosanthus grown on bunds has made a big difference in the milk yield of animals that helped in increased income. The farmers have also grown vegetables on the bund, which could partly meet the vegetable requirement of the family and have been possible due to moisture conservation, similar results were observed by Pandey et al. Similar results were obtained by maximum adoption (76.25%) was noticed for modern rice varieties followed by balanced fertilizer (64.71%).

4. *Adoption index of farmers as a result of intervention*

The information presented in Table 4 clearly reveals that majority (70.00%) of the farmers adopted *wadi* and one-third of the farmers were non-adopters. The mean adoption score of *wadi* adopters was 77.80. The mean difference of adoption score was 23.70, 'Z' value was found significant at 1 per cent level, which is due to the fact that the farmers from all the categories of wealth class had higher communication behaviour and consultancy pattern.

Table 4: Distribution of Farmers According to their Extent of Adoption

Category	Respondents				Adoption index	Z value
	Σf	Category	Σf	Percentage		
		Low	15	45.00	65.33	
Adoption	33	High	18	55.00	88.15	7.77*
		Overall	33	100.00	77.80	
Non-adoption	14	—	—	—	—	—

* Significant at 1% level.

5. *Correlation between knowledge and adoption*

A perusal of Table 5 indicates that there was a positive and significant relationship between the level of knowledge and extent of adoption in case of all wealth classes. On the other hand, it could be said that higher the level of knowledge, higher is the extent of adoption of *wadi* farmers.

Table 5: Correlation between Knowledge and Adoption

(N=33)

Category of wealth class	Correlation coefficient (r)
Very Poor	0.862**
Poor	0.775**
Medium	0.826**
Rich	0.370*
Very rich	0.612**
Overall	0.594**

** Significant at 1% level;

* Significant at 5% level.

Knowledge is an important factor influencing the adoption behaviour of farmers. While it is true that knowledge about the *wadi* practices alone cannot result in full adoption of the *wadi*, it is beyond any doubt. However, that higher knowledge about *wadi* acts as a strong internal motivating force for an individual in adopting *wadi*. Higher the knowledge of an individual, greater will be his/her efforts to acquire the needed hardware inputs in order to completely adopt the *wadi* practices.

Table 6: Constraints Perceived by the Non-adopters of *Wadi*

Issues	Constraints	Percentage
Economic related issues	The income is just sufficient to meet out the basic need—food, unable to pay wages for hired labours and previous financial stress due to bank loans	42.85
Social related issues	Commitment for higher education of children, cradle ceremony of children and commitment for marriage of daughters	50.00
Family related issues	Type of House (in terms of maintenance)—Small/Big, type of family—Nucleus/Joint family, family size—<5/>5 and kinship relations (the state of being related by blood) influenced by siblings/relatives	71.42
Peri-urban related issues	Commuting of male members to urban areas for work (Hubli), involvement of women in sale of curds in Hubli city, self-occupation of male members such as driver, skilled workers in city, etc. had made them not to involve in the *wadi* activity, due to Urban demands on flowers, vegetables and also immediate returns in short duration crops made him not to adopt *wadi* and men as well as women both are involved throughout the year in vegetable/fruit vending/flower trade in nearby city	85.71
Infrastructure related issues	Installation of electric poles/crossing wires/transformers hindering them not to adopt *wadi*	7.14
Natural resource related issues	Land: Small landholding (<1 Ac), land—leased out/leased in, joint holdings, legal disputes of land and land is more fertile Water: Water sources are far away from the field, non-availability of water throughout the year and protective irrigation is not possible	100
Human related issues	Youngsters are not interested in agriculture as they commute to city for work to get more wages, driving license holders, self-occupational skills and decision-making process in the family	35.71
Personal problems	If I adopt *wadi* I am unable to spend time in meetings and spending whole day outside in BAIF meetings, as I have lot of responsibilities at home such as taking care of shrewd bullocks, maintenance of milch animals and marketing of milk	21.42

Through participatory methodologies the farmers developed potentialities and gained knowledge to adopt *wadi* practices. The agricultural scientists provided better knowledge on *wadi* practices and by this they were confidant and knowledge support was with them throughout the project period. Similar results were obtained by maximum adoption (76.25%) was noticed for modern rice varieties followed by balanced fertilizer (64.71%).

The non-adopters expressed (100%) natural resource based issues such as small landholding (<1Ac), leased out land, leased in land, joint holdings, legal disputes of land and more fertile land (85.71%) scarcity of labour due to peri-urban related issues, more than 70 per cent family related issues (50%) social related issues, 42.85 per cent problem of inadequate finance, 42.85 per cent income problem (35.71%) human related issues and (7.14%) infrastructure related issues as the main constraints in adopting *wadi*.

These findings are inconsistent with the observation made by Shinde et al., 2003 who attributed that non-availability of sufficient irrigation water, shortage of capital, shotage of labour and non-availability of manures as some of constraints in adoption of recommended technology of groudnut cultivation. Gaikwad (1987); Khobragade (1988) and Rade et al. (1990) who attributed lack of knowledge about application of irrigation at critical stage and unavailability of sufficient irrigation water as main constaints in groudnut cultivation.

Conclusion

The motivational factors to adopt *wadi* practices were utilitarian motive, which ranked first followed by economic, achievement, affiliation and innovative motive. Participatory planning, implementation and consultancy pattern of community organiser and researchers were the most important activities throughout the project period to gain knowledge and adoption behaviour on *wadi* practices. The information on motivational pattern and adoption behaviour of *wadi* farmers would have a greater significance to focus on the future extension activities based on participatory approaches. It was also felt that the findings of the study would be of importance to local extension agencies, researchers and policy-makers and to extend their infrastructure to enhance adoption of improved practices of *wadi*.

REFERENCES

Dharminder, S. and Ravinder, K. (2004). Knowledge level of the Cotton Growers Regarding Integrated Pest Management Practices, *Journal of Extension Education*, 15:3611-18.

Gaikwad, A.B. (1987). A Study of Constraints in Adoption of Groundnut Production Technology in Sakri taluka of Dhule district of Maharashtra. *M.Sc. (Agri.) thesis,* MPKV, Rahuri, M.S.

Khobragade, H.G. (1988). Technological Gap between Recommended Technology and Actual Adoption by the Farmers cultivating Groundnut Crop. *M.Sc. (Agri.) thesis,* KKV, Dapoli, M.S.

Pandey, P.K., Sarkar, J.D., Sharma, M.L. and Suryawanshi, D.K. (2004). Constraints in Adoption of Recommended Rice Production Technology among the Farmers of Chattisgarh, *Journal of Extension Education*, 15:3633-38.

Rade, V.M., Dasi, B.R. and Girase, K.A. (1990). Constraints in Adoption of Recommended *Kharif* Groundnut Production Technology. Maharashtra *Journal of Extension Education*, 9:146-52.

Shinde, S.B., Bhoite, H.S. and Sawant, G.K. (2003). Constraints in Adoption of Recommended Technology of Groundnut Cultivation, *Journal of Maharashtra Agricultural Universities*, 28:62-66.

Vande Fliert (1998). "From Pest Control to Ecosystem Management: How IPM Training can Help". In: Dhaliwal, G.S., Randhawa, N.S., Arora, R. and Dhawan, A.K. (eds.) *Ecological Agriculture and Sustainable Development*, pp. 109-16. Centre for Research and Industrial Development, Chandigarh, India.

Problems and Constraints of Supplementary Enterprises in Dryland Integrated Farming Systems—A Case Study of Rabbitry in Himachal Pradesh

20

Ashok Kumar, K.C. Sood, Atul Dogra and *S.K. Sharma*

Introduction

Dryland farming refers to high-yield and high-efficient agricultural production in areas without irrigation capability and mainly depending on natural rainfall through the adoption of a set of dryland farming practices. It continues to be a key priority area for the Governemt of India, contributing nearly 44 per cent to the national food basket. However, income of farmers in these areas remained stagnant over the years due to low productivity and little marketable surplus.

The state of Himachal Pradesh has achieved the distinction of being regarded as model of hill development as agriculture and allied activities continues to be the mainstay of majority of the population, providing livelihood to about 71 per cent of the working population. But only about 17.2 per cent of net sown area is irrigated and the rest is all rainfed. In the state, agriculture exhibits the blend of both progressive and the traditional nature. The objectives of the government policy to meet immediate challenges of poverty, unemployment and malnutrition including full employment, economic growth, price stability, wage maintenance, protection of the national environment, greater opportunities for women and minorities. The effective route to development is to raise productivity of agricultural sector, which will then act as a stimulus to other areas of the economy. To boost the meagre income of small holdings there is a need for incorporation of the supplementary enterprises which not only increase productivity and income but also induces stability in the farm income. Rabbitry is one such micro-enterprise which is economical because of their early maturity, prolificacy (7-8 young ones) and short gestation period (31 days). The credit of introducing and popularising Angora rabbits in India goes to ICAR through its North Temperate Regional Station of Central Sheep and Wool Research Institute (CSWRI), Garsa, Kullu in Himachal Pradesh. The rabbits require a few inputs and can be reared by workers lacking in great physical strength. Rabbitry gives an opportunity to develop 20 subsidiary industries including nursery raising wool farming, cage fabrication; feed formulation, shearing, wool collection and grading, etc. The present study is an attempt to study the problems and constraints of rabbitry in Himachal Pradesh and suggest suitable policy measures.

Methodology

Complete enumeration of the rabbit farms in the state was done. The list of rabbit entrepreneurs in each district was obtained with the help of officials of the Development Blocks, progressive rabbit entrepreneurs, traders, concerned officials of government, etc. Though rabbitry for wool is taken-up in all districts except Una, Bilaspur and Lahaul and Spiti, where the climatic conditions are not congenial; Chamba, Hamirpur, Kangra, Kinnaur, Kullu, Mandi, Shimla and Solan districts were chosen for the study.

In each district a list of rabbit entrepreneurs was prepared along with their flock size. The rabbit farms were categorised into small, medium and large classes with the help of cumulative cube root frequency method according to distribution of the flock size. For each category the size and number of rabbit farms have been given in Table 1.

Table 1: Classification of the Rabbit Farms

Sr. No.	Category	Number of Rabbits	Number of Rabbit Farms	Percentage of Rabbit Farm
1.	Small	Less than 25	93	54.70
2.	Medium	25-150	45	26.47
3.	Large	$\geq$ 150	32	18.83
	Total		170	100.00

The problems and constraints confronted by rabbit entrepreneurs have been categorised into different sub-heads viz., production, financial, marketing and institutional problems. The percent response was recorded on different size of rabbit farms. To test the difference in the extent of problems confronted on different size of farms, chi-square test was employed to test the rejection or acceptance of the following null hypothesis:

H_o: pi 1 = pi 2 = pi 3 = pi 4

Pij = Problems on different categories of farms.

The value of chi square was computed as follows:

$$n\left[\frac{\Sigma k}{I=1}\ \frac{\Sigma k}{j=1}\ \frac{fij^2}{fionj} - 1 \sim\right] \chi^{2\,(k-1)\,(1-1)} df.$$

where

$$Fio = \frac{\Sigma l\ fij}{J=i} \qquad \text{(i.e. row total)}$$

$$n = \frac{\Sigma k}{i=1}\ fij \qquad \text{(i.e. column total)}$$

Results and Discussion

The problems and constraints faced by rabbit entrepreneurs have been categorised into four sub-heads viz., production, financial, marketing and institutional problems. The rabbit entrepreneurs were interviewed in respect of various problems and their response has been given in various sub-heads. The null-hypothesis (Hô: = pi1 = pi2 = pi3 = pi4), i.e. the extent of problems is same in all the categories of rabbit entrepreneurs was tested. The calculated value (χ^2 = 411.65) of Chi-square was greater than the table value (χ^2= 112.33). Therefore, it led to the rejection of null-hypothesis, i.e. hypothesis of no difference and revealed that the problems differ significantly in different categories and can be reduced if proper management practices are adopted by rabbit entrepreneurs and requires different way of tackling for each category.

Production problems of rabbit entrepreneurs

The production problems of rabbit entrepreneurs have been further sub-categorised into twelve types viz., availability of feed, price of feed, shortage of fodder, availability of cages, feeders, waterers, process of equipments, problem of pests and diseases, availability of veterinary aid, availability of medicine, availability of labour, high labour wages, availability of fresh blood (breeding animals to avoid inbreeding), maintenance of financial and pedigree records. The different problems confronted by wool rabbitry entrepreneurs are given in Table 2 and are as under.

The availability of quality feed is an important component in rabbit production. About 52 per cent of rabbit entrepreneurs reported that the feed was not available at right time. The rabbit entrepreneurs reported that the sub-standard feed led to deaths *en-masse* and heavy losses to them. The perusal of the table reveals that the gravity of the problem and farm size has an inverse relationship.

The feed prices when compared to output prices are quite high. About 72 per cent of the rabbit entrepreneurs on overall farms reported this problem and the gravity of the problem and farm size shows inverse relationship. The percent response decreased with increase in farm size. The commonly fed fodders in the study area are common grasses, bamboo leaves, *barseem*, *biul* leaves and kitchen waste. About 12 per cent of the rabbit entrepreneurs reported acute shortage of fodder in the study area especially during winter. The maximum shortage of fodder was experienced on medium farms (15.56 per cent) followed by small and large farms.

About 18 per cent of the entrepreneurs reported that the inputs like cages, feeders and waterers are not available. The extent of problem increased with the increase in farm size hence showed positive correlation between gravity of problem and farm size.

The problem of high prices of cages, feeders and waterers showed a positive correlation with the size of the farm. About 24 per cent of rabbit entrepreneurs on all farms reported the problems of high prices of these equipments.

Table 2: Production Problems of the Rabbit Entrepreneurs

(Percent)

Sr. No.	Particulars	Farm Size			
		Small	Medium	Large	All farms
1.	Quality feed not available at appropriate	72.04	35.56	16.13	52.07
2.	Feed prices not reasonable	94.62	51.11	35.48	72.19
3.	Shortage of fodder	11.83	15.56	9.68	12.43
4.	Cages, feeders and waterers not available at right time	6.45	28.89	35.48	17.75
5.	Price of cages, feeders and waterers are high	7.53	37.78	54.84	24.26
6.	Problem of pest and diseases	1.08	11.11	29.03	8.88
7.	Veterinary aid not available when required	7.53	35.56	45.16	21.89
8.	Medicines not available at right time	6.45	24.44	6.45	11.24
9.	Labour not available when required	58.06	22.22	16.13	40.83
10.	Labour charges not reasonable	37.63	26.66	25.81	31.95
11.	Non-availability of fresh blood	38.71	31.11	9.68	31.36
12.	Non-maintenance of financial and pedigree records	97.85	71.11	74.19	86.39
	Total number of farms	93	45	31	169

Note: Figures have been rounded off to their nearest digit.

The problem of pest and diseases was depicted by about 9 per cent of entrepreneurs on all farms and the percent response increased with the increase in farm size.

About 22 per cent of rabbit entrepreneurs confronted the problem of non-availability of veterinary aid on all farms. The gravity of non-availability of veterinary aid and farm size showed positive correlation. The maximum response of 45.16 per cent was reported on large farms followed by 35.56 per cent on medium farms and 7.53 per cent on small farms.

The non-availability of medicines was reported to the extent of about 11 per cent on all farms. The maximum percent response was recorded on medium farms 24.44 per cent. However, it was a tie (6.45 per cent) on small and large farms.

The non-availability of labour was experienced by about 41 per cent on all farms category. The gravity of the problem of non-availability of labour and farm size showed an inverse relationship. The response was maximum on small farms

(58.06 per cent) followed by medium farms 22.22 per cent on medium farms and 16.13 per cent on large farms.

The high wages of labour for attending rabbits was experienced by about 32 per cent of entrepreneurs on all farms category. The response recorded was as high as 37.63 per cent on small farms followed by 26.66 per cent on medium farms and 25.81 per cent on large farms.

The inbreeding in rabbits leads to decreased output, high incidence of diseases and mortality among the flock. Therefore, introduction of fresh blood of pure strain at frequent intervals is very important to eliminate the adverse consequences of inbreeding in a rabbitry. About 31 per cent of rabbit entrepreneurs on all farms reported the problem of non-availability of fresh blood. The gravity of the problem was maximum on small farms (38.71 per cent) followed by medium farms 31.11 per cent and 9.68 per cent on large farms thereby depicting inverse relationship between extent of problem and farm size.

The problem of maintaining financial and pedigree records was as high as about 86 per cent on all farms.

Financial problems of rabbit entrepreneurs

The establishment of rabbitry units requires some capital investment in the beginning. The finance is needed to construct shed, purchase cages, waterers, feederers and purchase of the breeding stock. The banks have a scheme to finance small and marginal farmers or agricultural labourers for small size units of 10 female + 2 male. However, if large entrepreneurs are interested to take up the scheme they could be financed more than one unit depending upon individual needs and merits of each case. The banks also make provision for providing veterinary aid, fodder and concentrated feed and training of beneficiaries. The loan limit is Rs. 2430 for a unit of 10 females + 2 males for fur/meat and Rs. 8850 for a unit of wool type breeding rabbits. The disbursement of loan is against purchase of equipments and breeding animals on the supply of bills of such items. The initial recurring expenses for first 6 months for pelleted feed have been capitalised term loan. The disbursement against supply of feed is also done against bills of supply. The loan against breeding animals is disbursed only when equipment and initial stock of one month feed is purchased and inexperienced farmer receives adequate training. Margin money by beneficiaries at the rate of 15 to 25 per cent for wool unit as per ARDC norms is envisaged. The rate of interest to ultimate borrowers will depend upon the amount of loan raised. However, the rate of interest varies from 12 per cent to 23.75 per cent with quarterly or half yearly interests. The security against the loan is hypothecation of rabbits, feed, fodder, equipment fixtures and other movables created out of the bank loan. Secondly, creation of equitable mortgage/charge over applicants' land where sheds are constructed and thirdly, third party guarantee of sufficient worth good for loan amount.

The supervision of the credit is through regular visits to the unit in order to ensure proper utilisation of loan amount disbursed for construction of sheds, purchase of equipments and rabbit stock. Thereafter, monthly inspections will be carried out regularly when the project commences.

The bank also makes a provision for adequate and comprehensive insurance cover in respect of shed equipment against fire, theft, earthquake, riots and rabbit mortality during the period of loan and relevant policies are also arranged in favour of bank. Banks assume that daily requirements of about two hours per unit for adult rabbits and the progeny, scheme would provide additional employment of 90 days per year per unit. The contacts with the farmers through survey revealed that majority of the farmers are not aware of the loan facility.

Table 3: Financial Problems of the Rabbit Entrepreneurs

(Percent)

Sr. No.	Particulars	Farm Size			
		Small	Medium	Large	All farms
1.	Credit not available at appropriate time	93.55	86.67	96.77	92.31
2.	Credit given not supervised	95.70	93.33	96.77	95.27
3.	Available credit is inadequate	93.55	86.67	96.77	92.31
	Total number of farms	93	45	31	169

Note: Figures have been rounded off to their nearest digit.

The non-availability of credit at appropriate time was reported by 92.31 per cent of rabbit entrepreneurs on all farms situation. The maximum response to this problem was recorded on large farms 96.77 per cent followed by small farms 93.55 per cent and medium farms 86.67 per cent. Banks advance loans only to those entrepreneurs whose risk bearing ability and repayment capacity is high. So small entrepreneurs do not gather courage to have loan under present circumstances when rabbit industry is passing through a slump.

About 95 per cent of rabbit entrepreneurs on all farms situations reported that credit once given is not properly supervised. The response recorded was maximum on large farms (96.77 per cent) followed by small and medium rabbit entrepreneurs respectively.

The rabbit entrepreneurs reported that credit given is not adequate 92.31 per cent of rabbit entrepreneurs on all farms reported that available credit was inadequate. Among different size of rabbit entrepreneurs, the maximum response was recorded on large farms and it was to the tune of 96.77 per cent followed by small farms 93.55 per cent and 86.67 per cent on medium farms.

Marketing problems

The availability of proper marketing facilities is a pre-requisite for the development of any enterprise. An efficient marketing system for the disposal of farm produce leads to faster growth and development. So far there is no stable and organised market for Angora rabbit farming on large scale. There is a demand for better breeds of Angoras but the sale/ disposal of ordinary surplus animals is not organised and ensured. The various marketing problems noticed on farms are shown in Table 4.

Transportation plays an important role in the marketing of produce which has to be shifted from the point of production to the point of final consumption. Comparatively the problem of transportation of the inputs and produce was revealed by a less number of rabbit entrepreneurs. On all farms situation only 11.24 per cent of entrepreneurs reported this problem. Maximum was recorded on large farms (45.16 per cent) followed by large farms (11.11 per cent). However, the problem of transportation was not confronted on small farms.

Table 4: Marketing Problems of the Rabbit Entrepreneurs

(Percent)

Sr. No.	Particulars	Farm Size			
		Small	Medium	Large	All farms
1.	Inadequate transportation facilities	0	11.11	45.16	11.24
2.	Regular markets for farm product	37.63	37.78	51.61	40.24
3.	Unremunerative price of meat	100.00	100.00	100.00	100.00
4.	Unremunerative price of wool	100.00	100.00	100.00	100.00
5.	Problem of grading of wool	29.03	2.22	9.68	18.34
6.	High marketing costs	10.75	31.11	61.29	24.26
7.	Support price by government	43.01	62.22	77.42	54.44
8.	Price of rabbits for remunerative	98.92	73.33	77.42	88.17
9.	Demand of wool products seasonal	100.00	95.56	100.00	98.82
	Total number of farms	93	45	31	169

Note: Figures have been rounded off to their nearest digit.

Rabbitry entrepreneurs in the study area reported to have insufficient market system. There were no ready buyers of rabbit products, especially wool, locally, so the farmers had either to transport it to distant towns or sell in the local market at low premium. About 40 per cent of the farmers on all farms situation responded to this problem.

The problem of unremunerative prices of meat was responded positively by each and every entrepreneur on all categories of farms. Rabbit meat could not be popularised among meat eaters in Himachal Pradesh. However, rabbit meat has many desirable traits like low fat and cholesterol, high proteins and minerals. The wool rabbits when become uneconomical with the passage of age are sold at throw away prices. Therefore, 100 per cent of rabbit entrepreneurs in all the categories confronted this problem.

The problem of unremunerative prices of wool has been responded positively by each and every rabbit entrepreneur on all the categories of farms. With the liberalisation the access to market by developed nations has increased. The Government of India has reduced tariff on import of wool to a greater extent and as a result the rabbit wool was available at a cheaper rate than home produced. Therefore, in the recent past year the prices of wool have slashed and frames are not getting remunerative prices for their produce.

The different grades of rabbit wool have different prices and traders cheat rabbit entrepreneurs by designating their produce as "B" grade wool. The problem of grading of wool before selling was revealed by 18.34 per cent of rabbit entrepreneurs on all farms situation.

The marketing costs in rabbitry include the cost of transportation, storage and other fees if any. High marketing costs were reported by 24.26 per cent of rabbit entrepreneurs on all farms situations. The problem of high marketing cost and size of the farm showed a positive correlation. The gravity of the problem of high marketing costs increased with the increase in farm size. On small farms 10.75 per cent of rabbit entrepreneurs responded to this problem. On medium farms (31.11 per cent) and on large farms 61.29 per cent reported this problem.

About 54 per cent of rabbit entrepreneurs on all farms situation revealed that the government should fix support price for rabbit wool. The per cent response of the rabbit entrepreneur increased with the increase in size of farm. The rabbits are purchased for establishing new rabbit units and also for cross-breeding purpose. About 88 per cent of rabbit entrepreneur on all farms situation revealed that the existing price of the animals is not remunerative. The problem of unremunerative prices was reported by 98.92 per cent entrepreneurs on small farms followed by large farms (77.42 per cent) and medium farms (73.33 per cent).

The demand for rabbit wool and other products is seasonal. During winter the demand for wool products rises, whereas, it decreases in other months of the year. The seasonal demand for wool products was reported by 98.82 per cent on all farms.

Institutional problems

The institutional problems faced by rabbit entrepreneurs have been studied under sub-heads like lack of training, extension facilities and availability of package of practices presented in Table 5.

The knowledge acquired through training is of great relevance. The different institutions organise training programmes for rabbit entrepreneurs. About 27 per cent of rabbit entrepreneurs on all farms situation reported that existing training facilities are not adequate.

The maximum response in favour of this problem was found on medium farms (51.11 per cent) followed by large farms (32.26 per cent) and small farms (12.90 per cent).

The rabbit entrepreneurs expect time to time help and guidance from different agencies of government. 84.62 per cent of rabbit entrepreneurs on all farms reported the lack of extension facilities. The improved technology developed by research institutes is disseminated through package of practices. 82.84 per cent of rabbit entrepreneurs reported the non-availability of package of practices. Cent percent response with regard to this problem was recorded on small farms. However, 53.33 per cent of medium farms and 74.19 per cent of large farms reported this problem, respectively.

Table 5: Institutional Problems of the Rabbit Entrepreneurs

(Percent)

Sr. No.	Particulars	Farm Size			
		Small	Medium	Large	All farms
1.	Training facilities not adequate	12.90	51.11	32.26	26.63
2.	Lack of extension facilities	93.55	66.67	83.87	84.62
3.	Package of practices available	100.00	53.33	74.19	82.84
	Total number of farms	93	45	31	169

Note: Figures have been rounded off to their nearest digit.

Future Strategy and Policy Implication

The rabbit wool industry in Himachal Pradesh is in recession. In order to revive the industry the following policy points may be taken into consideration by the planners and policymakers:

(i) For making rabbit as a viable livestock species it has to be introduced on scientific lines. The Government of India and Indian Council of Agricultural Research may open a National Rabbit Research Centre as has been done in case of sheep, goat, horses, camels, buffaloes and poultry thereby accepting rabbit as also one of the important livestock species having overall multifaceted utility to the country. Training of scientists, technicians and farmers should also be a component objective of such a National Rabbit Research Centre.

(ii) Case of slashing down of the import duty abruptly from 60 per cent to 15 per cent should be taken-up with the Ministry of Commerce, Government of India, so that the prices of the imported wool are at par with the domestic produce and domestic production is encouraged.

(iii) Good Angora rabbits are always in demand by the government agencies in other states which are keen to introduce rabbitry as supplementary enterprises. The rabbits are also required by private entrepreneurs within the state as well as in other states. The sale and disposal of surplus animals if organised through some state/ private marketing agencies/farmers' cooperatives will bring economic gains to Angora farmers.

(iv) The provision of training to rabbit entrepreneurs should be strengthened and package of practices should be readily available. Involvement and training of farm women has to be further encouraged. The farmers should make a co-operative and should shear the rabbit wool at one time for the uniform length and quality. Awareness among the entrepreneurs will abridge the gap between existing and recommended levels of inputs.

REFERENCES

Sharma, A.K., Mehta, P. and Sharma, S.K. (1998). "Economic Viability, Technological Gap Problems and Future Strategy to Enhance Export of Rabbit Wool in Himachal Pradesh." Seminar on Management of Export Marketing of High Value Farm Strategies, Hissar, pp. 152-60.

Bhat, T.K. (1999). "Angora Rabbit Farming in the Hilly Regions of India." *Himachal Veterinary Journal.* 3:61-63.

Cheeke, P.R., Patton, N.M. and Templeton, G.S. (1982). *Rabbit Production*, (5th edition), Published by the Interstate Printers and Publishers, INC, USA.

James, I. and M.C. Nitt (1986). "Waste Water in Rabbitries". *The Journal of Applied Rabbit Research*, Vol. 9, No. 2.

Jithendran, K.P. (1998). "Rabbit Production in Sub-Himalayan Region." *Indian Farming*. May 1998: 15-19.

Kumar, Ashok (2004). "Rabbit Wool Industry." Daya Publishing House.

Phull, Archana, (1991). "Economic Analysis of Rabbit Farming in Kangra District of Himachal Pradesh." M.Sc. Thesis (Unpublished), Himachal Pradesh Krishi Vishvavidyalaya, Palampur.

Sharma, A.K. and Sharma, S.K. (2001). Economic Analysis of Rabbit Farming Along with Technological Gap, Problems and Future Strategy. (Technical Lecture Copy): 135-41.

Sharma, A.K. (1998). "Untapped Potential of Rabbitry." *The Tribune*, December 5, 1998.

Sharma, A.K. (2000). "Rabbitry—A Hidden Treasure—Needs Attention." *The Himachal Reporter*, July 31, 2000.

21 Small Animals for Sustainable Farming—Selected Case Studies on Integrated Farming System

C. Cinthia Fernandaz and *K. Chandrakandan*

Case Study I–Diversifying Small Farms in Cambodia

A local NGO called Partnership for Development in Kampuchea (PADEK) helps through Self Help Groups (SHGs) and Agricultural Extension Network for sustainable agriculture, livestock and fisheries. Diversification, involving vegetable and fruits and live stock provides full employment to family even in dry season without depending on cities. This study implies "Choosing of more appropriate enterprise mix suitable for each locale for more employment and income."

- Animals kept in appropriate housing that is cleaned regularly.
- Animal manure used in biodigesters, fish-ponds or for compost.
- Animals vaccinated against common diseases.
- Animals fed a nutritious diet of locally available fodder.
- Return profits of greater than US$1.00 per labour-day.

Check-list for sustainability of fish production:

- Raising a diversity of fish species.
- Fish raised in pond that is integrated with the homegarden, rice fields.
- Pond has available water for more than 6 months of the year.
- Use locally available feed such as duckweed, rice bran and phytoplankton.
- Return profits of greater than US$1.00 per labour-day.

Case Study II—Crop-goat Integration

This was the experience of small farm holders of Vietnam. Growth and reproductive performance of the goats was of a high order in wet than dry season and the income increased, significantly. The results show that the project indicated that crop-small ruminant (goat) integration can be an important activity for small landholders of Asian region.

Development of the Cash Flow over the First Four Years of the Project (Average of the 10 Farms)

US$

	2001	2002	2003	2004
Returns				
Cash income				
Crops	580	862	1425	1400
Goat kids	450	675	900	1012
Total cash income (1)	1030	1537	2325	2412
Value of non-cash income				
Added value of livestock	656	875	1093	1312
Manure	94	100	100	120
Total non-cash income	750	975	1193	1432
Total gross returns	**1780**	**2512**	**3518**	**3844**
Costs				
Supplies	304	170	180	180
Goat house	125	20	20	20
Loan interest	65	53	41	29
Loan repayment	200	200	200	200
Total cash costs (E)	694	443	441	429
Non-cash costs				
Labour	1200	1200	1200	1200
Total non-cash costs	1200	1200	1200	1200
Total costs	**1894**	**1643**	**1641**	**1629**
Net profit	–114.00	869.00	1877.00	2215.00
Net cash return (I-E)	336	1094	1884	1983

Case Study III—Sustaining Livelihoods through Small Animals

Valasaguttapalli is a village in the drought prone area of Madanapalli Taluk at Chittur distict in the state of Andra Pradesh. During 2000, Chaitnayee an NGO identified as eco-network partners of AMEF programme and mobilised eco-member group. Here, sheep was preferred over goats as it can graze on dry grass, twigs and crop residues. These animals could give a fresh life to the villages. The group could attract a bank loan to the tune of Rs. 1 lakh. Thus, Valasaguttapalli has become model village.

Case Study IV—Rearing Practices of Small Animals in Himalayas

Small animals have been a source of livelihood to the tribals of Lahaul Nalky. The sheeps and goats were folked together. Labour was shared and indigenous knowledge and wisdom alone were adopted.

Case Study V—Chicken-bamboo Farming in Southern China

The chicken-bamboo system is evolving and there is still room for improvement. To improve marketing of chickens, the government's support is needed.

Case Study VI—Improving Livelihood through Vermi-composting

Vermi-composting making through SHG's is a good example and strategy by combining technical interventions with community mobilisation, the rural poors have been brought out of poverty.

Case Study VII

Pig Rearing—A Means to Food Security

The mountain region of Darjeeling district is characterised by weak marketing infrastructure, high levels of emigration, etc. Pig farms enthused many farmers for rearing cross breeds without any extra investments.

Case Study VIII

Mr. Devarajan of Burgur, Dharmapuri district, Tamil Nadu had added apiculture as an enterprise in his integrated farming system. He had been keeping five beehives in an acre of land, which was given by the government free of cost. Queen bee also was given by the department. He collected source bee comb from the nearby forest. He placed queen bee in it. The total cost of the enterprise worked out to be Rs. 992/5 boxes/ac and the cost:benefit ratio was 1:1.5877.

Case Study IX

This is Mr. Venkateshbabu, N. from Burgur, Dharmapuri district, Tamil Nadu, who has added silvipasture in his farm. His annual income was about 6 lakh. He had been doing silvipasture in ten acres of land. Summer ploughing was done and sowing of the various fodder species had been taken place under the mango trees. He used to sow about 25 kg of Cenchrus, 25 kg of Stylosanthus and horse gram during the monsoon period of June-July. The total cost of cultivation worked out to be Rs. 5700/ha and the return realised was in the ratio of 1:1.728.

Case Study X

Mr. A. Thangavelu of Idapadi, Salem district has added dairy as an enterprise, along with conventional crops. The dairy was maintained in a more scientific way.

When there was vegetation they allow animals for natural grazing and in dry season he uses dry fodder for feeding. He provides concentrates regularly. He used to feed the animal by green fodder @ 15 kg/day (or) dry fodder @ 5 kg/day besides 25 kg of concentrates per month. The total production cost was Rs. 6418.00/animal/year and the return realised was Rs. 1.4475 per rupee.

Case Study XI

This is Mr. Ganapathy of Vennandur, Namakkal district, Tamil Nadu. He had poultry in one acre of farming. He had 10,000 birds, divided into 4 batches @ 2500 birds/batch. He prepared own feed by using the ingredients like sorghum, maize, ragi, dryfish meal, groundnut oil, minerals and antibiotics. The total cost of production calculated was Rs. 38048.25/100 birds. The cost: benefit ratio was 1:1.5445.

Case Study XII

Mr. Mathalaimuthu of Burgur Dharmapuri district has practiced sericulture along with crops. He raised mulberry in 0.5 acre of land. Training had been provided by the Department of Sericulture. He underwent training before starting the practice. Inputs including mulberry cuttings were brought from the

Department of Sericulture. Rearing was maintained by application of correct leaves depending upon the stage of larvae. Total cost of production was Rs. 11,850/ha/2 crops per year and the return was Rs. 1.68 per one rupee investment.

Conclusion

Small animals like sheeps, goats, rabbits, pigs, local chickens, etc. are always better adaptable and easily manageable by the dryland small farm holders in addition to others viz., agri-horticultural crops. This kind of farming systems approach have been found to be highly remunerative and generates more employment throughout the year for both family and hired labourers and avoids emigration to cities.

REFERENCES

1. LEISA INDIA Newsletter, September 2005.
2. Mary Elizabeth (2001). "Integrated Dry Farming Systems in Tamil Nadu—A feasibility Study." Unpublished M.Sc (Ag) Thesis, Department of Agricultural Extension and Rural Sociology, TNAU, Coimbatore-3.

Integrated Farming System (IFS) with Special Focus on Organic Farming: Approaches and Possibilities in the Context of Indian Agriculture

22

N. Deepa, A. Rohini and *S. Moghana Lavanya*

Introduction

Different farming systems are prevailing in India primarily due to varied agricultural base, available resources and location specific needs of humans, animals, etc. Farmers through their experience have established this farming system to meet their food, fuel and fibre requirement in a manner that they are least dependent on the external source. However, advancement in crop production technology and need for higher foodgrain production compelled the nation for adoption of crop based production system. Integrated farming system, which lacked database for further innovations, was marginalised in the wake of green revolution. Increasing consciousness about conservation of environment as well as health hazards associated with agro-chemicals and consumers' preference to safe and hazard-free food are the major factors that lead to the growing interest in alternate forms of agriculture in the world. Organic agriculture is one among the broad spectrum of production methods that are supportive of the environment. The demand for organic food is steadily increasing both in the developed and developing countries with an annual average growth rate of 20-25%. Organic agriculture, without doubt, is one of the fastest growing sectors of agricultural production.

Waterlogging and secondary salinisation have been the banes associated with excess and irrational irrigation. Groundwater table declined sharply as more and deeper borewells were drilled.

Recharging of groundwater has also been reduced due to severe deforestation. Indiscriminate use of chemical pesticides to control various insect pests and diseases over the years has destroyed many naturally occurring effective biological control agents. An increase in resistance of insect pests to chemical pesticides has also been noticed.

Health hazards associated with intensive modern agriculture such as pesticides residues in food products and groundwater contamination are matter of concern. The occurrence of multi-nutrient deficiencies and overall decline in the productive capacity of the soil due to non-judicious fertilizer use has been widely reported. Such concerns and problems posed by modern-day agriculture gave birth to new concepts in farming, such as organic farming, natural farming, biodynamic agriculture, do-nothing agriculture, eco-farming, etc. The essential feature of such farming practices imply, i.e. back to nature.

The Concept of Organic Agriculture

Organic agriculture is one among the broad spectrum of production methods that are supportive of the environment. Organic production systems are based on specific standards precisely formulated for food production and aim at achieving agro-ecosystems, which are socially and ecologically sustainable. It is based on minimising the use of external inputs through use of on-farm resources efficiently compared to industrial agriculture. Thus, the use of synthetic fertilizers and pesticides is avoided.

Organic Farming and Country's Food Security

The primary concern of all organised communities and civilised societies is to meet the food requirements of its people. The cultivated area, required to maintain the present level of foodgrain production in India without using the fertilizers, reaches more than the total geographical area of the country. At present, there is a gap of nearly 10 million tonnes between annual addition and removal of nutrients by crops which is met by mining nutrients from soil. A negative balance of about 8 Mt of NPK is foreseen in 2020, even if we continue to use chemical fertilizers, maintaining present growth rates of production and consumption. The most optimistic estimates at present show that only about 25-30% nutrient needs of Indian Agriculture can be met by utilising various organic sources. It is proved beyond doubt that on long-term basis, conjoint application of inorganic fertilizers along with various organic sources is capable of sustaining higher crop productivity, improving soil quality and soil productivity. The organic sources should be used in integration with chemical fertilizers to narrow down the gap between addition and removal of nutrients by crops as well as to sustain soil quality and to achieve higher crop productivity. The food security demand of the country requires that inorganic fertilizers be used in balanced doses.

Issues

Organic farming has the twin objective of the system being sustainable and environmentally benign. The Indian farmer should get the advantage of emerging global market on organic farming which is at present around 26 billion US dollars and is expected to grow to 102 billion US dollars in 2010. Currently, 130 countries are producing certified organic products. A vast scope for promotion of organic farming in the export market, without compromising with the national food security exists in the country, as farming by tribals and under rainfed conditions is generally organic since very little chemical inputs are used. Organic farming as a concept/philosophy is well tested in some of the Western countries though the same is not unknown to most of the nations.

But in the Indian context, it needs to be looked into more critically seeking answers to the following questions: What level of crop productivity is acceptable? Is it suitable for a country like India with such a large population to feed or can it fit in the niche areas? Are available organic sources of plant nutrients sufficient for organic farming in the form it is advocated? Are organic farming technologies sustainable in the long run? Redefining the concept of organic farming in Indian context, major recommendations are listed below:

Organic farming: Prospects and limitations

- For India with its ever increasing population, the sustainable agriculture has to be based on site-specific balanced and adequate fertilization and an integrated plant nutrient supply system (IPNS) involving organics, inorganic and bio-fertilizers. What the country needs today is the conjunctive use of organic and inorganic sources of plant nutrients for sustainable productivity.
- Organic farming is a market demand driven agriculture aimed to cater to the foreign export and affluent section of the society in the country. However, in order to make a dent in the export market, we need to develop high-tech organic technology with strict quality control meeting international quality standards prescribed for organic produce.
- Niches of the organic farming need to be identified. However, the real niches will be determined by the market infrastructure and the international links. The practice should be considered for lesser endowed regions of the country rather than in resource endowed regions which serve as the backbone of the country's food security.
- The availability of organic manures in adequate amounts and at costs affordable by the farmers is a major problem. The increased mechanisation has further reduced the availability of manures with the farmers and this problem will become more acute in future. In such circumstances, post-harvest residues should be utilised to the fullest extent. However, to accomplish this objective, feasible technologies are needed for *in situ*.
- Recycling/rapid composting of on-farm residues and wastes, in addition to extension efforts to change the mindset of the farmers. Possibilities of using non-traditional organic sources e.g., slaughter house waste, should be exploited to partly supplement plant nutrient needs of the organic farming systems.

- Changing cropping patterns with area under legumes is going down, shrinking area under green manures due to economic considerations and reduced availability of loppings from forests seriously restricts wide scale use of green manures. Inclusion of legumes in intensive cereal production systems as short duration grain or forage crops, as substitute to one of the cereals or as break crops needs to be promoted.

Organic farming: Its relevance to Indian agriculture

- In Indian context, organic farming has to be practiced without synthetic pesticides, but complete exclusion of fertilizers may not be advisable under all situations. A holistic approach involving integrated nutrient management (INM), integrated pest management (IPM), enhanced input use efficiency and adoption of region-specific promising cropping systems would be the best organic farming strategy for India.
- As organic farming is attracting worldwide attention and there is a potential for export of organic agricultural produce, this opportunity has to be tapped with adequate safeguards so that the interest of small and marginal farmers is not harmed.
- Organic farming may be practiced in crops, commodities and regions where the country has comparative advantage. To begin with the practice of organic farming should be for low volume, high value crops like spices, medicinal plants, etc. beside fruits and vegetables, for which Research and Developmet (R&D) support is required.
- Organic farming should not be confined to the age old practice of using cattle dung and other inputs of organic/ biological origin, but an emphasis needs to be laid on the soil and crop management practices that enhance the population and efficiency of belowground soil biodiversity to improve nutrient availability. Performance of cultural techniques for weed control and that of bio-pesticides for pest management need to

be evaluated under field conditions, preferably under cultivator's management conditions. Besides the identification of regions suitable for the adoption of organic farming, the crops and their products should also be identified which are amenable for production through organic ways and have the potential to fetch a premium price in the international organic market.

Organic farmers and NGOs

- Region-specific resource inventory including animal wealth, farm residues/by-products and their competitive uses, non-conventional nutrient sources of organic/biological origin, etc. has to be prepared for development of rational technology packages of organic farming.
- A strong technological back-up by scientific community should be provided in order to verify, confirm and further refine some selected ITKs like Agnohotra, Panchgavya pertaining to organic farming.
- Crop-specific and farming situation-specific package of practices for organic cultivation should be developed and after thorough on-farm validation, recommended for adoption. Such proven technology packages need to be documented in regional languages.
- Entrepreneurial potential with respect to production of organic inputs, processing and marketing of organic food should be fully exploited.

Trade and certification issues

- Certification of organic produce is an important issue that is central to organic production itself as the prime goal of organic farming is and will be to fetch premium price in domestic and international market for the producers.
- For desired economic gains out of organic farming, the upcoming global market on organic produce has to be exploited, for which strict phytosanitary measures have to be followed.

- A strong research back-up has to be put in place to develop and improve national standards for organic farming. The policy documents brought out so far by APEDA and DAC including the report of Task Force on Organic Farming, etc. should be considered in developing such scientifically sound standards, mainly for the crops and commodities that have export potential. The national standards should be the same for domestic market and for export.
- Weak links in the certification systems have to be identified and researchable areas flagged.
- Whereas organic certification will continue to be a process certification, strong research set-up/laboratories are required to monitor the quality of organic produce so as to prevent the sale of sub-standard material in the name of organic produce and to save the interests of producers and consumers.
- Organic produce fetches premium price owing to better quality and for credibility in the market it ought to be quality goods. Therefore, there should be adequate provision for their grading, packaging, storage and transportation. Marketing outlets on the lines of milk unions may be established.
- There is a need to organise Producers Cooperative Marketing Societies and establish credible marketing channels for steady flow of organic foods and materials in accordance with the demand, as also to safeguard the interest of small farmers opting for organic agriculture.

The quality aspects of organic food

There is an urgent need to compare the quality of organically produced food with conventionally produced food. There appears to be a widespread perception amongst consumers that organically produced foods are of superior nutritional quality. However, to prove or disapprove this contention very limited research has been conducted and whatever meager scientific data is available is often outdated or based on inadequate study designs lacking proper controls. In view of this, the following points merit consideration:

- No clear-cut evidence is available to support consumer perceptions regarding potential health benefits of organic foods. An in-depth research on quality aspects is required to arrive at any valid acceptable conclusion.
- Valid nutritional quality comparisons between organic and conventional food requires that plants be cultivated in similar soils, under identical climatic conditions, be sampled at the same time, pre-treated similarly and analysed by validated methods.
- Well-designed controlled studies in animal models and human subjects are needed.
- It is also necessary to undertake well-controlled studies to evaluate sensory properties, shelf-life and nutrient load of organic produce vis-à-vis produce from the conventional farming techniques.
- More importantly, organically and conventionally produced food should also be analysed for pesticide residues and microbiological safety, i.e. presence of pathogenic organisms which could pose health hazards.

Priority research areas

- Develop package of practices for integrated pest management for organic farming in different agro-climatic regions for specific crops involving components like bio-inoculants, pheromones, etc. among others.
- It is frequently documented that fertilizers and pesticides applied at recommended rates have had no adverse effect on soil biological activity and that integrated farming systems are best for nutrient management, yield sustainability and soil biodiversity conservation. There is thus a need to develop modern organic farming system integrating the best available options.
- There is a need to establish referral laboratories for analysis of pesticides, heavy metals and mycotoxins in the produce with appropriate accreditation to help

organic farming movement. The maximum residue limits in organic food must be set in accordance with the CODEX standards.

- The Green Revolution Technologies have been alleged to cause depletion of soil organic carbon. The critical values of soil organic matter that can support the sustainable crop production under organic management have to be worked out.
- The patterns of rate and amount of nutrients released from various organic sources and their goodness of fit with the nutrient requirement of the crops at different growth stages need to be worked out. This information could be used in evolving appropriate nutrient management schedule, so as to ensure optimal nutrient supply to the crop at active physiological stages having peak nutrient demands.
- Allelopathic effects of various plant species need to be tapped particularly for weed and pest management.

Conclusion

The interest in organic agriculture in developing countries is growing because it requires less financial input and places more reliance on the natural and human resources available. Studies to date seem to indicate that organic agriculture offers comparative advantage in areas with less rainfall and relatively low natural and soil fertility levels. Labour is almost non-existent. Organic agriculture does not need costly investments in irrigation, energy and external inputs but rather organic agricultural policies have the potential to improve local food security especially in marginal areas. Possibly, the greatest impact of organic agriculture is on the mindset of people.

The following conclusions can be drawn on important issues regarding organic farming:

- Organic farming is a market demand driven agriculture aimed to cater to the foreign export and affluent section of the society in the country. However, in order to make a dent in the export market, we need to develop high-tech organic technology with strict quality control meeting international quality standards prescribed for organic produce.

- Organic farming may be practiced in crops, commodities and regions where the country has comparative advantage. To begin with the practice of organic farming should be for low volume high value crops like spices, medicinal plants, etc. beside fruits and vegetables for which R&D support is required.
- There is a need to organise Producers Cooperative Marketing Societies and establish credible marketing channels for steady flow of organic foods and materials in accordance with the demand, as also to safeguard the interest of small farmers opting for organic agriculture.
- Organic foods are proved superior in terms of health and safety but there is no scientific evidence to prove their superiority in terms of taste and nutrition as most of the studies are often inconclusive.
- There is an urgent need to compare the quality of organically produced food with conventionally produced food.
- Integrated farming systems are best for nutrient management, yield sustainability and soil bio-diversity conservation. There is thus a need to develop modern organic farming system integrating the best available options.

REFERENCES

FAO (1999). Guidelines for the production, processing, labeling and marketing of organically produced foods. Joint FAO/WHO Food Standards Program Codex Alimentarius Commission, Rome, CAC/GL 32, p. 49.

Kler, D.S., Kumar, A., Chhinna, G.S., Kaur, R. and Uppal, R.S. (2001). Essentials of organic farming—A review. *Environ. Ecol.*, 19, pp. 776-98.

Kortbech-Olesen, R. (2000). Export opportunities of organic food from developing countries. In *World Organics*, Agra Europa (London) Ltd., London, UK, 9-10 May 2000.

National Academy of Agricultural Sciences (NAAS) (2005). "Organic Farming: Approaches and Possibilities in the Context of Indian Agriculture." Policy Paper No. 30, p. 8.

Sellen, D., Tollamn, J.H., McLeod, D., Weersink, A. and Yiridoe, E. (1993). In *The Economics of Organic and Conventional Horticulture Production.*, Department of Agricultural Economics and Business, University of Guelph, Canada.

Veeresh, G.K. "Organic farming: Ecologically sound and economically sustainable." Paper presented at the International Conference on *Ecological Agriculture: Towards Sustainable Agriculture,* Punjab Agricultural University, 15-17, November, 1997.

Constraints and Prospects of Farming in the Hills of Himachal Pradesh

23

K.C. Sood, K. Bassi, S.S. Masand, Ashok Kumar and *Atul Dogra*

ABSTRACT

Himachal Pradesh is a hill state having about 84 per cent small and marginal farmers. In the state, the average size of landholdings is only 1.16 hectare and more than 82 per cent of the cultivated area is rainfed. The land productivity in the state is low due to small landholdings, rainfed farming, poor soil characteristics, subsistence farming, difficult input supply and low rate of technology adoption. Most of the cultivated area in the state is under cereals viz. wheat, maize, rice and barley which are less remunerative.

The diverse agro-climates in the state provide a comparative advantage of growing various crops suitable to these environments only. However, the intensive cultivation of various high-value crop enterprises could mainly be limited only to the 18 per cent irrigated area and in about 82 per pent rainfed areas still subsistence farming is being done.

The present study is based mainly on secondary data along with field experience. The state is demarcated into four agro-climatic zones. In the lower hills the rainfed areas are suited for food crops like maize and wheat, fodder crops like sorghum, oilseed like sesame and brassica pulses like black-gram and chick-pea along with fruits like mango and citrus. However, in the irrigated areas, rice, sugarcane, wheat, *barseem*, vegetable crops and fruits like mango, litchi, citrus, guava, dairying, apiculture, mushroom farming, etc. are profitable. The mid-hills (irrigated and rainfed) are suitable for growing vegetables, ginger and turmeric along with cereal crops, tea, pecan-nut, mushroom farming, bee-keeping and dairying particularly crossbred cows.

The high hills are more suited for the cultivation of vegetable crops like french-bean, peas, cabbage and cauliflower along with fruits like apple, pear and apricot, etc. However, the farming in the hills is subjected to some major unavoidable constrains like small landholdings, rainfed agriculture, subsistence farming and lack of technological empowerment of women farmers and remoteness and inaccessibility which result in unsustainability and low productivity of the farming systems.

In order to make the rainfed agriculture, sustainable, eco-friendly and profitable, the agriculture production technology like run-off harvesting and its recycling, increasing the retentive capacity of soil, *in situ* water conservation, increasing water use efficiency, contingent planning and intercropping, etc. are very important and need to be advocated on larger scale. The suggested diversification plan for the entire state, if included in the agricultural development plan of the state, can help in improving the socio-economic status of the farmers to a greater extent.

Introduction

Himachal Pradesh is a hill state located in western Himalayas having different agro-ecological zones. The state has mostly small and marginal farmers in widely scattered and remote villages. It has by and large remained backward and traditional in agriculture mainly due to small landholdings, rainfed agriculture, subsistence farming and due to lack of technological empowerment of women who otherwise play a dominant role in agriculture. However, the variable agro-climates in the state provide a comparative advantage of growing various crops suitable to these environments only, i.e. tea, off-season vegetables, variety of fruits, flowers, seed potato, quality seed production of European vegetables, medicinal and aromatic plants, apiculture, rabbitry and some other commercial crops are grown profitably in some niche areas, providing good livelihoods to the farmers of the state. Animal husbandry and forestry are also very important in the state.

The State of Himachal Pradesh has achieved the distinction of being regarded as a model of hill development. In Himachal

Pradesh, the average size of landholding is only 1.16 ha and more than 84 per cent holdings are small and marginal. The soils of the state are degraded and fragile having shallow depth. The lands are undulated, sloppy and characterised by terraced farming. More than 82 per cent of the cultivated area is rainfed.

Rainfall is the most important water resource in the state. There are 12 districts out of which 9 receive annual rainfall of around 1000 mm or more. Agriculture is practiced under rainfed conditions as majority of districts have around 80 or more cropped area as rainfed. The average annual rainfall for the entire state is 1032 mm. In spite of this huge rainwater potential, the crops suffer for want of irrigation water during sowing and some critical growth periods. Quite often there is a complete failure of the crop even for want of timely pre-sowing irrigation. The major reason for this situation is the erratic behaviour and uneven distribution of rains. About 80 per cent of the rain in the first three agro-climatic zones is received from mid-June to mid-September. Hence, sowing of both *rabi* and *kharif* crops in the absence of pre-sowing irrigation is delayed.

Most of the cultivated area in the state at present is under cereals, viz., wheat, maize, rice and barley. However, the food-grain crops alone cannot sustain the family in the times to come. Agriculture and allied activities continue to be the mainstay of majority of the population as they provide livelihood to about 71 per cent of the working population and contribute about 22 per cent of total income of the state.

To make the hill-agriculture economically viable, eco-friendly and sustainable, diversification in terms of crops and crop-related supplementary enterprises including livestock husbandry holds great potential. Jodha (1986) also stressed that emphasis should be laid on total farming system rather than individual crops for sustainable farming system. Divedi and Awasthi (1987) observed that the intercropping system for rainfed lands reduced the risk of total crop failures, thereby, stabilising the income. Himachal Pradesh has highly diverse and variable agro-climates and this diversity offers opportunities to grow a range of off-season vegetables, fruits, flowers, medicinal, aromatic and dye (MADPs) plants, quality seed production of temperate vegetables and mushroom cultivation along with dairying, apiculture, rabbitry, cold water fisheries and sheep-goat rearing. However, the intensive cultivation of various high-value crop enterprises could mainly be limited only to the 18 per cent irrigated area and in about 82 per cent rainfed areas still subsistence farming is being done thus making it economically unviable.

Dryland farming refers to high-yield and high-efficient agricultural production without irrigation capability and mainly depending upon rains through the adoption of a set of dryland farming practices, suitable to these areas. Since in the country vast cultivated land is under rainfed agriculture (about 65%), adoption of dryland farming practices/strategies is the key priority area and this is true for the state of Himachal Pradesh as well, where most cultivated area is rainfed and the prime concern is to improve the productivity of agriculture under such conditions.

The present investigation is an attempt—

(i) To examine the various problems and constraints of agro-climatic zones of Himachal Pradesh

(ii) To study various problems and constraints of hill agriculture

(iii) To investigate the prospects of hill farming both under irrigated and rainfed conditions and

(iv) To suggest appropriate strategies for rainfed agriculture.

Methodology

The present study is based upon the both primary and secondary data and field experience. The secondary data have been obtained from various publications of the Department of Economics and Statistics, Government of Himachal Pradesh and publications of the CSK Himachal Pradesh Agricultural University and other related publications. To examine the various problems and constraints of the different agro-climatic zones, farmers have been interviewed from time to time.

Results and Discussion

The results of the study have been summarised under sub-heads:

Agro-climatic zones of Himachal Pradesh

Himachal Pradesh has highly diverse and variable agro-climates ranging from sub-tropical to alpine and the state can be clearly demarcated into four agro-climatic zones as given in Table 1 and districtwise coverage of the different zones has been given in Table 2 and the coverage of area has also been shown in Fig. 1.

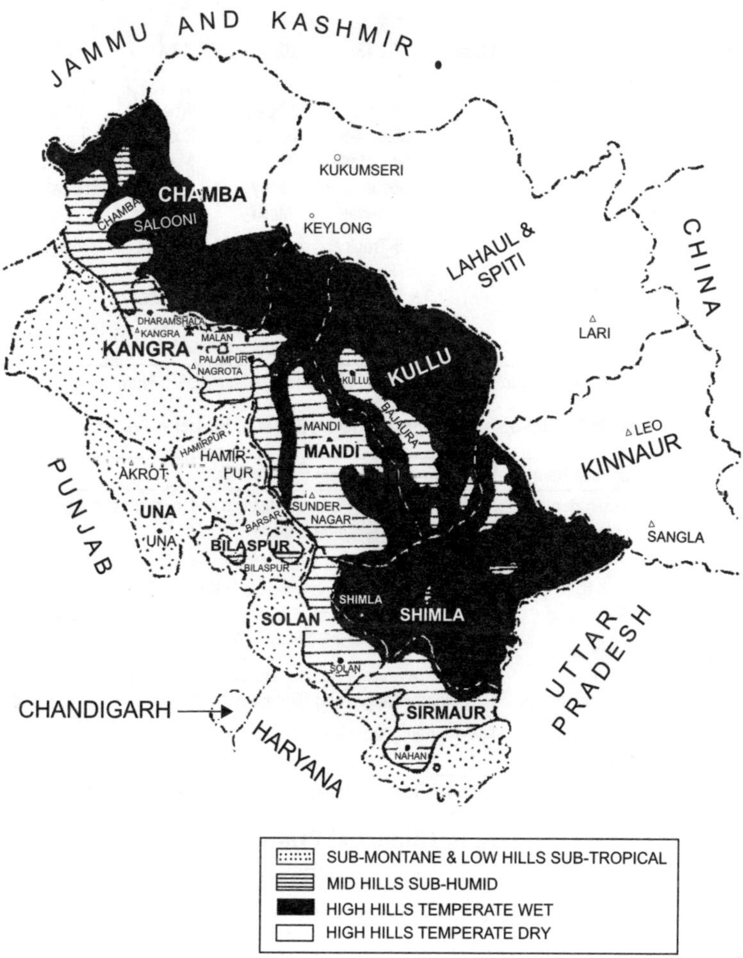

Fig. 1: Agro-climatic Zones of Himachal Pradesh

Table 1: Salient Features of Agro-Climatic Zones of Himachal Pradesh

Sr. No.	Particulars	Zone-I	Zone-II	Zone-III	Zone-IV
1.	Total geographical area ('000 ha) (%)	913.2 (16.4)	1183.2 (21.3)	1280.9 (23.0)	2190.0 (39.3)
2.	Elevation (m above m.s.l.)	Below 650	650-1800	1800-2200	> 2200
3.	Raiiifall (mm)	1000	1500-3000	1000	250
4.	Total cropped area ('000 ha)	355.10	383.40	171.80	24.00
5.	Percent irrigated area to total cultivated area	16.60	17.50	7.80	40.60
6.	Percent cropped area of total geographical area	32.43	32.40	13.41	1.10
7.	Major field crops	Wheat, Maize, Rice, Gram, Mash, Sesame, Brassicas	Wheat, Maize, Rice, Barley, Mash	Wheat, Potato, Maize, Barley, Rajmash	Barley, Potato, Pea, Buckwheat
8.	Major Fruits	Sub-Tropical Fruits	Sub-Temperate Fruits, Mango, Litchi	Apple, other temperate Fruits, Nuts	Nuts, Dry Fruits, Apple

Source: National Agril. Research Project Report of Himachal Pradesh, ICAR, New Delhi, 1982.

Table 2: District-wise Coverage in different Zones of Himachal Pradesh

Zones	Special Name	Coverage
Zone-I	Sub-mountain Low-hills Sub-tropical	Una, Bilaspur, Hamirpur districts and parts of Sirmaur, Kangra, Solan and Chamba districts
Zone-II	Mid–hills Sub-humid	Palampur and Kangra tehsils of Kangra district. Rampur tehsil of Shimla and parts of Mandi, Solan, Kullu, Chamba and Sirmaur districts
Zone-III	High-hills temperate-wet	Shimla (except Rampur tehsil) parts of Kullu, Solan, Chamba, Mandi, Kangra and Sirmaur districts
Zone-IV	High-hills temperate dry	Kinnaur, Lahaul-Spiti districts and parts of Chamba district

Source: National Agril. Research Project Report of Himachal Pradesh, ICAR, New Delhi, 1982.

Constraints of farming in Himachal Pradesh

(i) *Small and marginal farmers*: The average size of holdings is 1.16 ha and more than 84 per cent of the holdings are small and marginal as shown in Table 3. The small and scattered holdings discourage the farmers from adoption of improved production technologies, thus, leading to subsistence farming.

Table 3: Distribution of Land holdings—Marginal and Small Farmers Hold the Key

Size of holding (ha)	Category of farmers	No. of holdings (lakh)	Area (lakh ha)	Average size of holding
Below 1.0	Marginal	5.56 (64.4%)	2.30 (23.0%)	0.4
1.0-2.0	Small	1.73 (20.1%)	2.41 (24.1 %)	1.4
2.0-4.0	Semi-medium	0.95 (11.0%)	2.56 (25.6%)	2.7
4.0-10.0	Medium	0.34 (3.9%)	1.94 (19.5%)	5.7
10.0 and above	Large	0.05 (0.6%)	0.78 (7.8%)	15.6
Total		8.63	9.99	1.2

Source: Economic Survey Himachal Pradesh, 2004-05. Economics & Statistics Department , Govt. of Himachal Pradesh, Shimla.

(ii) *Rainfed agriculture*: More than 82 per cent of the cultivated area is rainfed. Although high precipitation is received in the state, mainly during rainy season, yet the rainfall is uncertain and unevenly distributed, thus, adversely affecting productivity, especially of *rabi* crops. Rather, the high amount of rains received during monsoon results in excessive run-off and soil losses.

(iii) *Degraded and inherent deficient and fragile soils*: The soils of the state are degraded and fragile having shallow depth. The lands are undulated, slopy and characterised by terraced farming and, as such, these inherently deficient soils are not suitable for growing those crops which are grown more profitably in the plains.

(iv) *Dominant role of women in agriculture but lack of participation in decision making*: In the state, most of the farm operations are carried out by the women. Role of the women folk in decision making in the NE hills is much greater than that of the NW hills, including Himachal Pradesh, who need technological empowerment to play a greater role in decision making.

(v) *Inaccessibility and remoteness*: Because of difficult terrain, most of the hills are inaccessible and approachable on foot only and this remoteness creates hurdles in dissemination of improved technologies, low rate of technology adoption, difficult input supply and marketing problems for agricultural produce.

Because of these general features of the hill farming, the state by and large has remained traditional and backward in agriculture. Nevertheless, hilly and mountainous areas have potential of growing high-value cash crops, fruits, flowers and off-season vegetables.

Prospectus of profitable farming in the hills

Diverse and variable agro-climatic conditions in the hills offer a unique opportunity to grow a range of off-season vegetables, fruits, flowers, quality pulses and MADPs, besides traditional crops, along with taking up of supplementary enterprises. The diversification of farming in the state is proving beneficial to the farmers, turning subsistence farming into a commercially viable venture. A number of success stories on account of diversification in different niche areas is also visible.

The state of Himachal Pradesh has earned a name in temperate fruit production particularly apple, the dominating fruit crop of the state, which accounts for about 41.6 per cent of the total area under fruit crops and 76.8 per cent of the total fruit production. As such, the state is known as the 'Apple State' of the country. The area under apple which was about 400 ha only in 1950-51 increased to 92,820 ha in 2001-02. The production of different fruits over 5 years is given in Table 4.

Table 4: Fruit Production in Himachal Pradesh

('000t)

Particular	2000-01	2001-02	2002-03	2003-04	2004-05
Apple	376.73	180.53	348.26	459.49	527.60
Other temperate fruits	20.45	29.24	63.02	40.65	60.16
Nuts and dry fruits	2.75	2.91	3.26	3.57	3.73
Citrus	11.06	20.46	16.03	28.12	25.84
Other sub-tropical fruits	17.04	30.31	29.05	28.14	67.71
Total	428.03	263.45	459.62	559.97	685.04

Source: Economic Survey 2004-05. Directorate of Economics and Statistics, Government of Himachal Pradesh, Shimla.

Similarly, mango has emerged as an important fruit crop in the low-hill areas (parts of Bilaspur, Hamirpur, Una and Kangra districts) in the recent years. Citrus and litchi are the other potential fruit crops of the lower hills adjoining the plains.

The state has also earned a name in the off-season vegetable production. These vegetables have put the farm economy on a sound footing with an annual turnover of about Rs. 750 crore. The area and production of principal vegetable crops grown in Himachal Pradesh are given in Table 5. The progressive farmers of the state who have taken up the diversification of agriculture, along with any of the supplementary enterprises like mushroom cultivation, dairying, apiculture, rabbitry, cold-water fisheries and sheep-goat rearing as per their capabilities and climatic suitability, are earning decent livelihoods.

There has not been an appreciable change in area under crops as a result of diversification if we look at the cropping pattern changes in the state through the years 1972-73 to 1999-2000 (Table 6, Kumar et al., 2002 b). This is because of the fact that diversification has been restricted to certain specific niche areas in districts like Solan, Kullu, Shimla, Sirmour, Lahaul and Spiti and Kinnaur.

Table 5: Area and Production of Principal Vegetable Crops in HP (2004-05)

Vegetable Crops	Area (ha)	Production (Mt)
Garden Pea	14,789	1,54,043
Tomato	9,013	2,46,033
Cabbage	2,790	71,134
Cauliflower	1,648	29,328
Beans	2,444	27,202
Capsicum and Chillies	1,642	17,862
Cucurbits	1,993	49,830
Radish, Turnip and Carrot	1,222	20,157
Onion and Garlic	2,725	45,835
Brinjal	544	10,070
Bhindi	1,373	13,774
Potato	15,000*	1,70,000*
Ginger	4,000*	—
Other vegetables	2,829	46,082

*Estimated area and production.

Table 6: Cropping Pattern Changes in HP—1972-73 to 1999-2000 (as %age of Total Cropped Area)

Crop	1972-73	1982-83	1992-93	1995-96	1999-2000
Wheat	34.27	39.24	38.93	37.66	38.73
Maize	28.16	30.10	31.90	32.58	31.35
Paddy	10.54	9.45	8.42	8.74	8.38
Minor millets	10.45	7.34	4.87	4.75	4.31
Pulses	7.76	4.48	4.22	3.80	3.40
Oilseeds	2.58	2.43	2.25	2.19	1.97
Total vegetables	2.11	2.22	2.61	3.15	3.69
Potato	1.56	1.39	1.46	1.44	1.53
Other vegetable	0.55	0.83	1.15	1.71	2.16
Ginger	0.24	0.20	0.17	0.14	0.21
Other cash crops	1.10	0.88	0.82	0.98	0.98

Existing Agricultural Scenario in Different Zones

The different existing farming systems under rainfed and irrigated farming along with their gross and net returns in different zones have been presented in Table 7 (modified from Kumar et al., 2003).

Table 7: Different Existing Production Systems in Zone—I, II and III of Himachal Pradesh

Farming Systems	Small Farmers					
	Zone I		Zone II		Zone III	
	Gross Returns (Rs/ha)	Net Returns (Rs/ha)	Gross Returns (Rs/ha)	Net Returns (Rs/ha)	Gross Returns (Rs/ha)	Net Returns (Rs/ha)
A. Rainfed Farming (Rs/ha)						
1. Maize - Wheat	7,980	5,170	8,197	5,637	8,176	5,699
2. Maize + Pulses - Wheat	8,460	5,633	8,821	6,153	–	–
3. Maize - Wheat + Sarson	8,057	5,190	–	–	–	–
4. Maize + Til-Wheat + Gram	9,193	6,246	–	–	–	–
5. Maize - Toria- wheat	9,971	6,372	10,682	6,507	–	–
6. Maize - Gram	6,323	3,672	–	–	–	–
7. Til - Gram	8,603	4,623	–	–	–	–
8. Paddy - Wheat	5,865	3,415	5,465	3,198	–	–
9. Paddy - Fallow	2,014	565	2,181	692	–	–
10. Maize + Soybean - Wheat	–	–	9,188	6,394	9,152	6,399
11. Maize - Peas	–	–	12,195	8,548	12,410	8,654

12. Paddy - Barley	–	–	3,694	1,705	–	–
13. Maize - Barley + Sarson	–	–	6,898	4,928	–	–
14. Maize - Fallow - potato - wheat (biennial)	–	–	–	–	24,142	12,836
15. Maize + pulses - Fallow - Potato - Wheat (biennial)	–	–	–	–	25,042	13,550
16. Maize - Fallow - Potato - Peas (biennial)	–	–	–	–	28,156	15,751
B. Irrigated farming (Rs/ ha)						
1. Paddy - Wheat	10,628	7,496	9,850	6,651	8,040	4,863
2. Paddy - Barseem	11,029	8,924	10,289	7,624	–	–
3. Chari - Barseem	13,438	11,993	–	–	–	–
4. Brinjal - cabbage - potato	49,114	36,006	–	–	–	–
5. Cauliflower - radish - potato	51,863	38,929	–	–	–	–
6. Cauliflower - Turnip - wheat	39,786	36,154	–	–	–	–
7. Potato - Capsicum - Spinach - Cauliflower	71,464	57,028	–	–	–	–
8. Tomato - Bottlegourd - Peas	63,864	54,954	–	–	–	–
9. Sugarcane - Ratoon	14,229	9,389	–	–	–	–
10. Maize - Potato - Potato	–	–	45,402	31,175	–	–
11. Cauliflower - Frenchbean - Cauliflower	–	–	68,783	60,425	–	–
12. Tomato - Radish - Peas	–	–	90,727	85,194	–	–
13. Tomato - Wheat	–	–	68,373	63,653	–	–
14. Tomato - Tomato - Cauliflower	–	–	1,06,652	96,406	–	–
15. Tomato - Tomato - Peas	–	–	99,152	90,594	–	–
16. Brinjal - Cabbage - Cauliflower	–	–	48,956	40,547	–	–
17. French bean - French bean - Peas	–	–	–	–	90,255	81,802
18. Cabbage - Peas	–	–	–	–	41,251	35,686
19. Cauliflower - Peas	–	–	–	–	72,249	64,678
C. Dairy Milch Animals (Rs/ animal)						
1. Cross-bred cow	7,078	3,148	8,102	4,287	7,038	3,728
2. Buffalo	6,281	2,789	6,915	3,172	5,541	2,509
3. Local Cow	1,894	251	2,975	1,148	2,006	6,67
D. Fruits (Rs/ha)	18,400	14,516	22,777	18,547	33,227	31,371

Note: These three zones account for more than 97 per cent of cultivated area.

Table 8: Gross Farm Income from Different Enterprises in Zone I, II and III of Himachal Pradesh

(Percent)

Enterprises	Small Farmers		
	Zone I	Zone II	Zone III
Field crops including cereals, pulses, oilseeds, fodder crops and sugarcane	34.23 (90.74)	31.79 (90.36)	14.97 (74.88)
Vegetables	12.28 (8.38)	23.10 (9.64)	7.66 (Other Veg.) + 15.76 (Potato) (7.09) (18.03)
Fruits	1.48 (0.88)	6.32	30.68
Farm Forestry	0.88	1.38	0.90
Dairy	43.05	31.49	22.00
Other livestock*	8.08	5.92	8.03
Total income over which the percentages have been worked out (Rs)	14,882 (100)	18,383 (100)	20,581 (100)

*Includes income for sheep/goats/mules/ponies; poultry, bullock hire out services and sale of animals, value of wool, value of dung.

Note: Figures in parentheses indicate the percentage of area under different crop groups.

The data depict that the farming systems which involved diversification with vegetable crops are highly remunerative and among different vegetables, Frenchbean, Cabbage, Cauliflower and Peas are more important under irrigated conditions in Zone III, whereas Tomato, Cauliflower, Radish, Peas and French beans give higher returns as compared with other vegetables under irrigated conditions in Zone II and in Zone I, under irrigated farming, Potato, Capsicum, Spinach, Cauliflower, Bottlegourd and Peas give higher returns. The returns from different crop groups and agriculture-related activities are given in Table 8.

The perusal of Tables 7 and 8 (Kumar et al., 2002 a and 2003) shows that field crops including cereals, pulses, oilseeds and fodder crops play a vital role in the farm economy in Zone I and Zone II. However, dairying also contributes significantly in Zone I. The income from vegetable crops is conspicuous in Zone II, whereas the share of fruits and potato was appreciable in Zone III. This segregation of income from different enterprises provides vital statistics for designing the future strategy for diversification in the hills.

Constraints of rainfed farming

The socio-economic status of the farmers of rainfed areas is generally low because of the following constraints:

- Farming is risky and unprofitable.
- Cultivation of high value crops, vegetable, off-season vegetable not possible.
- Very little scope of extensive agriculture.
- Meager employment generation.
- Limited scope of agro-based industry.
- Choice of crops is very limited.
- Full potential of crops cannot be harvested.
- Timely application of package of recommendation is not ensured.
- Increased cost of cultivation.
- Minimal mechanisation is possible.
- Quality of the produce is not up to the mark.
- Better livestock rearing is not feasible.
- Overall income is low.
- Feasibility of agro-based supplementary enterprises is low.

Strategy for rainfed farming

The income of the farmers in rainfed areas remained stagnant over the years due to low productivity. In order to increase the income and to improve the socio-economic status of the farmers, the following agricultural production strategy should be implemented in rainfed areas:

Run-off harvesting and recycling: Water harvesting is one of the most promising and technically feasible recommendations where rainfall is high but erratic and most of which is lost as run-off. In order to increase the land productivity in rainfed areas, run-off harvesting and its recycling is one of the most important components. Since the majority of the farmers have small landholdings, the rainwater storage in small dug-out silpaulin lined tanks, may be for inter-terrace water harvesting, serves more useful purpose for the individual farmer (Acharya and Kapur, 1997). Check dams must be constructed from the upper to lower ridges on a watershed basis for recharging of wells and tube-wells and minimisation of soil erosion through the reduction in the velocity of the flowing water.

Increasing retentive capacity of soils: The hilly soils are generally exposed to erosion which leads to variation in soil characteristics, low retentivity of water and nutrients, huge run-off losses and ultimately low crop yields. The use of organic manures, adoption of soil conservation measures and cultivation as per land capability classification will improve the retentive capacity of these soils.

In situ rainwater conservation: In rainfed areas, if residual rainwater is conserved *in situ* and carried for sowing of either wheat or maize, the timely germination of the crop can be ensured and latter rains are enough to take care of the crops. Soil moisture conservation through mulching with organic wastes like pine needles, lantana, eupatorium and grasses have been found effective in rainfed areas (Acharya et al., 1998). The horticultural crops and improved grasses can be grown successfully in rainfed areas if rainwater is conserved in contour trenches.

Diversification in agriculture: In rainfed areas, the socio-economic status of the farmers can be improved by growing vegetable and fruit crops, maintenance of better livestock, pisci-culture, bee-keeping, poultry, rabbitary and floriculture. Acharya and Kapur (2000) have highlighted the importance of shifting from conventional maize-wheat cropping sequences to vegetable based cropping system, if water is recycled from rainwater harvesting tanks.

Rejuvenation of hill springs: Due to continuous pressure on natural resources, the hill springs are getting dried up. The catchment of hill springs should be rejuvenated to increase the discharge rate which can further be stored in tanks and recycled for profitable use.

Water harvesting in cold deserts: An indigenous method of water storage called "Zing" has been developed by the farmers of Spiti valley for raising crops under hostile climates (–40°C to +35°C), Choudhary Sarvan Kumar Himachal Pradesh °Krishi Vishvavidyalaya, Palampur has constructed silpaulin lined water harvesting tank to store snow melt water for growing crops.

Increasing water use efficiency: The water use efficiency under all cropping systems should be the main objective of rainfed areas through adoption of science based crop technology including use of high yielding varieties, integrated nutrient management and other agronomic manipulations. The drip and sprinkler irrigation systems can play a key role in achieving high levels of productivity per unit of water consumed.

Contingent Farming: In hilly regions, the rains are erratic and in most of the cases the sowing of crops is delayed. Before the adoption of any cropping systems, contingent planning should be given due consideration. For example, if sowing of maize is delayed due to late monsoon rains, farmers should go for the sowing of late sown short duration crops suitable for the areas.

Intercropping: Under rainfed conditions, in order to avoid risk of crop failure, intercropping with low water requirement crops is a viable proposition. Wheat can be intercropped with gram and mustard whereas maize with horse gram, soybean and sesamum, etc. in Zone I and Zone II of the state. Divedi and Awasthi (1987) also reported that intercropping may be followed for rainfed lands to reduce the risk for crop failures, thereby, stabilising the income of the farmers.

Short Duration Crops: The maximum benefit of stored rain-water can be achieved if stored rainwater is recycled for short duration crops. As a long-term strategy, early maturing and drought tolerant varieties of different crops need to be developed to harvest the maximum benefit out of limited water availability.

Cultivation of selected medicinal, aromatic and horticultural plants: Some selected medicinal, aromatic and horticultural plants can be grown successfully under rainfed conditions. These crops have enough scope to improve the socio-economic status of the farmers. However, for growing these crops, technical expertise as well as proper market tie up is also required.

Diversification Plan for Different Areas of Himachal Pradesh

Based on the available primary and secondary data, field experience and success stories, the authors suggest a crop and related supplementary enterprises diversification plan as outlined in Table 9.

The suggested diversification plan, if included in the agriculture development plan of the state in a comprehensive way, can contribute greatly in the socio-economic upliftment of the farmers in different niche areas of the Himachal Pradesh.

Table 9: Suggested Diversification Plan for Zone I-IV of HP

Zone	Farming situation	Proposed cropping system	Potential thrust area	Supplementary enterprises
I	Rainfed	Maize + Sesame-Wheat + Gram/Mustard Maize-Toria-Wheat	Rainfed tomato (transplanted in 1st week of August) Colocasia in niche areas having quality advantage of the produce, Mango, Amla	Dairy (Buffalo), Poultry, Apiculture, Mushroom cultivation
	Irrigated	Maize fodder+ Cowpea fodder -Oat+ Sarson fodder Potato Capsicum-Spinach-Cauliflower Potato-Capsicum-Early pea-Cauliflower Tomato-Bottlegourd-Pea Capsicum-Cucumber-Spinach/Radish Maize+Cucumber Radish-Radish-Peas Maize+Ginger-Radish-Garlic Horticulture: Mango Citrus, Litchi, Strawberry	Selected vegetable-based cropping pattern Mango, Litchi and Strawberry Selected ornamented, medicinal, aromatic and dye crops	
II	Rainfed	Maize-Toria-Wheat Maize+Soybean-Wheat Maize-Pea-Radish	Pecan nut Pomegranate, stone fruits	Cross-bred cows, Apiculture,

Zone	Farming situation	Proposed cropping system	Potential thrust area	Supplementary enterprises
			Resource-poor farmers can take up cultivation of flowers like marigold Selected aromatic, medicinal and dye crops	Mushroom cultivation
	Irrigated	Tomato-Tomato-Cauliflower Tomato-tomato-Pea Brinjal-Cabbage-Cauliflower Brinjal-Cauliflower-Cauliflower	Diversification to vegetables like Tomato, Brinjal, Cabbage, Cauliflower Resource-rich farmers could opt for the cultivation of flowers like Gladioli, Lilium, Carnation, etc. Kiwi, Strawberry	
III	Rainfed	Maize-Peas Maize-Fallow-Potato-Wheat	Diversification with vegetable crops Seed production of temperate vegetable crops Apple, Pomegranate, Walnut, Apricot	Cross-bred cows and Mushroom cultivation Rabbitry, cold-water Fishery
	Irrigated	Tomato-Cabbage/Cauliflower-Pea/Radish Apple, Pomegranate and other temperate fruits Floriculture	Organic farming of vegetables Selected ornamental, medicinal, aromatic and dye crops	
IV	Irrigated	Seed Potato (Lahaul valley) Pea, Cabbage, Cauliflower, Capsicum Hort.: Apple Dry fruits (Apricot, Almond, Chilgoza, Resingrapes, Walnut)	Organic vgetables Seed Production of temperate vegetables Flower Seed Production Organic fruits High-value crops e.g. Kalazira and Saffron in Sangla valley of Kinnaur Hops, Kuth in L&S Selected medicinal and aromatic crops Seabuckthorn	Sheep and goat rearing Yak x Cow hybrids

REFERENCES

Acharya, C.L. and Kapur, O.C. (1997). Water storage in sloping lands. Low cost input and sustainable agriculture (ILEIA), Newsletter Netherlands 13:30.

Acharya, C.L. and Kapur, O.C. (2000). "Management of soil and water resources for sustainable development of agriculture in low and mid-Shivaliks of H.P." 50 Years of Research on Sustainable Resource Management in Shivaliks. Ed. S.P. Mittal, R.K. Aggarwal and J.S. Samra, CSWCRTI, Research Centre, Chandigarh, 417-30.

Acharya, C.L., Kapur, O.C. and Dixit, S.P. (1998). Moisture conservation for rainfed Wheat production with alternate mulches and conservation tillage in hills of N.W. India. *Soil and Tillage Research* 46:153-63.

Divedi, R.N. and Awasthi, R.P. (1987). Intercropping of soybean in maize under rainfed agro-climatic conditions of Nagaland. *Seed & Farmers* 13 (2): 18-21.

Jodha, H. (1986). Research and technology for dry farming in India—Some issues for Future strategy. Indian *Journal of Agricultural Economics* 41 (3): 234-37.

Kumar, A., Vashist, G.D. and Sharma, S.K. (2003). Sustainable production systems: Implications of nutritional security and quality of life in Himalayan Region. Research Volume on *Quality of Life in Himalayan Region.*

Kumar, Ashok; Sharma, S.K. and Vashist, G.D. (2002 a). Profitablity, risk and diversification in mountain agriculture: Some policy issues for slow growth crops. *Indian Journal of Agricultural Economics* 57 (3): 365-66.

Kumar, V., Sharma, R.K. and Sharma, H.R. (2002 b). Changing cropping in Himachal Pradesh—A district-wise study. *Himachal Journal of Agricultural Research* 28 (1+2): 58-62.

Integrated Farming System—Means to Achieve Sustainable Economic Returns for Small and Medium Farmers

24

D.S.M. Gowda, S.M. Hiremath, P.V. Kusagur and *K.B. Yadahalli*

ABSTRACT

Ten villages of Haveri district were selected using PRA techniques, under consultancy services for designing, laying and conducting farming system studies for rainfed agro-ecosystem through Sujala Watershed Project. The prime objective of the study was to identify the biotic and abiotic factors influencing the agronomic practices followed under particular farming system. Appraisal revealed that small and marginal farmers were very much inclined towards the cultivation of commercial crops, which, however, involve high degree of risk and uncertainties. These cropping practices had lead to injudicious use of available farm resources, alongside enforcing the marginal farmers to bring costly external inputs into the land. To overcome the associated risks and uncertainties modified sustainable farming system was suggested which mainly emphasised inclusion of all the enterprises.

Before adoption of integrated farming system the farmers had a cropping pattern with cereals and pulses (64.28%), commercial crops (21.42%) and vegetables (14.30%). The cropping pattern in the modified system was cereals and pulses (68.93%), oilseeds and vegetables (10.68% each) and commercial crops (9.71%). The modified sustainable farming system had helped the small and medium farmers to meet family and animal needs. The sale of eggs, aged birds and milk ensured in generating income throughout the year.

Introduction

Majority of the farmers of Haveri district have very small landholding. Mono-culture is caught in a web of difficulties particularly forced upon under unevenly distributed rainfall in the 70% of rainfed/dryland of the state and there is degradation of soil, loss of soil and depletion of major and micro-nutrients (Siddaramappa, 1996). The income from the mono-cropping is hardly sufficient to sustain the farmers' family with gradual decline in farm size, it has become increasingly difficult to produce enough food for the family for normal living. Due to this fact most of the farmers go to cities in search of a job. Under such circumstances, to ensure a regular income for decent living (above poverty line) farmers have to undertake some land based enterprises which will complement existing farming activities to get more income leading to social and economic upliftment.

Integrated farming system is one of the potential approach, which emphasis a judicious combination of any one or more of such enterprises and effective recycling of residues/wastes for better management of available resources with the small and marginal farmers for generation of more income and employment to the family members during off season. It also, ensures the recycling of the available resources/by-products within the farm itself. With cropping system approach, better crop coverage, utilisation of natural biological forces and better management through legume, cereals, grasses, fodder, vegetables and fruit/ tree species besides soil nutrient exploitation is minimised, productivity is improved and diversified needs of family are met (Sharma et al., 1991).

With this background to achieve stable farm income, integrated farming system activities were undertaken with following objectives:

(i) Meeting the diversified needs of the house

(ii) Maintaining/improving production and

(iii) Less dependence on external inputs such as chemical fertilizers and plant protection chemicals.

Methodology

The objectives of integrated farming system were studied by selecting forty eight farmers from ten villages of Haveri district. Before actual implementation of integrated farming system, multi-disciplinary core team was formulated to develop suitable farm modules for the given micro farming system. For this purpose to know the farmers' needs and farming perspective necessary information was obtained through participatory rural appraisal (PRA). Under this climate, soil, cropping pattern and requirement of the farmers were studied and analysed by following simple tabular analysis method in order to arrive at meaningful conclusion to suggest farm plan modules to the small (< 1 ha) and medium (> 1 ha and < 2 ha) farmers. Two farmers from each category were selected from ten villages.

The farming study was started from *kharif* and *rabi*/summer season of 2004-05. Further to study the impact of suggested farm modules for different categories of farmers were compared and evaluated for their feasibility with benchmark information like land use pattern and cost and return realised from the respective modules. After adoption of IFS activities, all recommended packages of practices were followed in each aspect of the farming system. In addition to the crops, dairy, poultry and vermi-compost units were also included in this farming system.

Results and Discussion

Efficiency of utilisation of land and other resources increased owing to adoption of IFS by the farmers as given in Table 1. Increase in land use was to the extent of 47.14 per cent owing to introduction of enterprises such as oilseeds, vegetables, poultry and vermi-composting units. The cropping intensity also increased due to introduction of horticulture components. Further various crops taken up as intercrop and relay crops also attributed to such large increase in the area utilisation. Though the percent area under the commercial crops decreased, their was compensatory increase in the area under the oilseeds and food grains. Further the area, which was unutilised hitherto, was used for planting of castor and fodder grass. This facilitated maximal

Table 1: Crop Enterprise Analysis for the Small and Medium Farmers before and after the Adoption of Integrated Farming System

Sl. No.	Particulars	Before adoption of IFS (acres)	After adoption of IFS* (acres)
1.	Food grains	4.50 (64.28)	7.10 (68.93)
2.	Commercial crops	1.50 (21.42)	1.00 (9.71)
3.	Oil seeds	—	1.10** (10.68)
4.	Vegetables	1.00 (14.30)	1.10 (10.68)
Intensity of land use		7.00	10.30
Percentage increase in land use		—	47.14

* With fruit crops ** Includes castor on bunds.
Values in parentheses indicate percentage.

Table 2: Returns Realised and Costs Incurred by the Small and Medium Farmers before and after the adoption of Integrated Farming System+

Sl. No.	Particulars	Before adoption of IFS			After adoption of IFS*		
		Total returns (Rs.)	Total cost (Rs.)	Net returns (Rs.)	Total returns (Rs.)	Total cost (Rs.)	Net returns (Rs.)
1.	Food grains	8000 (54.98)	4800 (57.14)	3200 (52.03)	16000 (40.61)	5600 (42.91)	10400 (39.47)
2.	Commercial crops	3000 (20.62)	1900 (22.62)	1100 (17.89)	4500 (11.42)	1600 (12.26)	2900 (11.01)
3.	Oilseeds	—	—	—	5400 (13.71)	2550 (19.54)	2850 (10.82)
4.	Vegetables	3550 (24.40)	1700 (20.24)	1850 (30.08)	7000 (17.77)	2000 (15.33)	5000 (18.98)
5.	Poultry: sale of eggs/ aged birds	—	—	—	2500 (6.35)	500 (3.83)	2000 (7.59)
6.	Vermi-compost	—	—	—	4000 (10.15)	800 (6.13)	3200 (12.14)
	Total	14550	8400	6150	39400	13050	26350
Percentage increase over		—	—	—	170.79	55.36	328.46
Cost benefit ratio		—	—	1.73	—	—	3.02

* From fourth year onwards fruit yield about Rs. 5000 worth fruits.
Values in parentheses indicate percentage.

use of land, harvesting soil moisture, in addition to build-up of resilience of soil owing to waste incorporation from crop husbandry and animal husbandry. Further increase in components from of crop husbandry also helped to increase soil bioflora.

Adoption of IFS by the small and medium farmers has its economic impact. IFS is advocated especially to the above farming groups with an objective of sustained income. The details of comparative monetary returns from various enterprises are compiled in Table 2.

Before adoption of IFS bulk of the returns was from food grains whose market rate comparatively low in comparison to commercial crop and vegetables. However, after introduction of various enterprises there was uniform distribution of income from all the enterprises. The worth noticing feature is steady income to the farmer all through the year from either of enterprises adopted. C : B ratio increased from Rs. 1.73 to Rs. 3.02 for every rupees spent.

Conclusion

The approach of integration of farming system is practical, profitable, acceptable and help each other maintaining ecological balance to the farmers in the rainfed areas for maximum utilisation of natural resources in the farm area. Thus, the risk and uncertainty encountered due to the biotic and abiotic factors have been reduced for sustainable yield.

Acknowledgement

Authors are thankful to the Commissioner, Watershed Development Department, Sujala Watershed Project, Bangalore, for financial assistance for conduct of integrated farming system demonstrations.

REFERENCES

Siddaramappa, R. (1996). "Integrated Nutrient Management, Organic Farming and Sustainable Agriculture," Proceedings of the National Seminar held at UAS, Bangalore, India. October, 9-11, pp. 90-101.

Sharma, L.R., Bhati, J.P. and Ranveer Singh (1991). Emerging Farming System in Himachal Pradesh—Key Issues in Sustainability. *Indian Journal of Agricultural Economics*, 46 (3): 422-28.

25

IFS Demonstrations—An Approach for Sustainable Development

S.M. Hiremath, D.S.M. Gowda, P.V. Kusagur, H.R. Nagaraju and *R.A. Budihal*

ABSTRACT

The crop husbandary activity in Haveri district of north Karnataka under rainfed situation is mainly dependent upon south-west mansoon. The average rainfall ranges from 450 to 650 mm per annum with least cropping period. Monoculture is a prominent cropping system observed in this area. Owing to unpredictable weather conditions, crop failure and unstable market prices small and marginal farmers experience lot of problems. Due to these perpetual problems satisfactory returns to the investments have been not realised and growers are exposed to high degrees of risk and uncertanity. Under such rainfed culture to ensure sustainable living, adoption of integrated farming system practice is of great significance. Based on the agro-climatic conditions and socio-economic status of farmers integrated farming system demonstrations were undertaken under consultancy services for designing, laying and conducting farming system studies under rainfed agro-ecosystem through Sujala Watershed Projects by Krishi Vigyan Kendra, Hanumanamatti, in Haveri district by judicious mix of agriculture enterprises like dairy, poultry, fruit crops, vegetables, vermi-compost, etc. All recommended packages of practices are followed in each aspect of the farming enterprises. Analysis of the first year data regarding economics revealed that the results are encouraging and farmers will definitely get higher returns than their conventional cropping system (CCS). Further integrated farming system will earn more profit when fruit crops will start bearing. The benefit to cost ratio increased from 0.61 to 1.56. The additional advantage being in terms of employment generation to the extent of 275 man days from 180 man days in the conventional cropping system.

Introduction

In Haveri district, more than 61 per cent of the total farming community belongs to marginal and small farmers having only 29.42 per cent of the total operational area. Agriculture is predominant economic activity in the district which contributes 40 per cent of the total income (Anon., 2005). The conventional cropping system, sole cropping with gradual decline in farm size and occurrence of three consecutive drought years (2000-01, 2001-02 and 2002-03) it has become increasingly difficult to produce enough food and other farm produces for the family day-to-day maintenance. Under these circumstances, to ensure a regular income farmers have to undertake added land based enterprises. In this regard IFS is one of the potential approach for the small and marginal farmers by combining more than one enterprises in their farm to increase the productivity and supplement the income. Hence, to generate income and employment opportunity for the farmers of Haveri district by integrating different land based enterprises IFS demonstration were undertaken.

Methodology

Based on the objectives of IFS, demonstration were conducted purely under rainfed culture in the farm of marginal farmer in Kamdod village of Haveri district in North Karnataka was started during *kharif* and *rabi*/summer 2004-05 and it was continued for 2005-06. Involving field crops, vegetables, fruits, spices, dairy, poultry and vermi-compost enterprises. All recommended package of practice were followed in each aspect of the farming system.

Before actual implementation of demonstrations multi-disciplinary core team of KVK, formulated farm modules as per the farmer needs and resources. Further impact and suggested farm modules of farmer was compared and evaluated for their feasibility with benchmark information. The data of conventional cropping system and after adoption of IFS, were analysed by following simple tabular analysis technique to arrive at meaningful conclusion. The component combination and area of allocation under different enterprises/activities in the farm are presented in Table 1.

Table 1: Analysis of IFS Demonstration in Respect of Sri. B.S. Melamalagi, Kamadod Village in Ranebennur Taluka of Haveri District

Sl. No.	Particulars	Conventional Cropping System (2003-04)		After adoption of Integrated Farming System (2004-05)	
I	Landholdings (acres)				
	Food grains		7-00		4-00
	Commercial crops		0-20		1-00
	Oilseeds		0-00		1-10
	Vegetables		0-00		1-10
	Total		7-20		7-20
II	Crops Grown	Sole crops:	Greengram (local) Maize (pvt. hy.) Sorghum (local) Cotton (local)	Sole crops:	Cotton (DHH-11), Sunflower (RSFH-1), Tomato (Megha) clusterbean (Pusa sadabahar)
				Intercrops:	Greengram (Sel-4)+ R. gram (Maruti) 4:2 Blackgram (TAU-1)+ R. gram (Maruti) 3:1 Navane (HMT-100-1) + R. gram (Maruti) 4:2
				Fruit crops:	Sapota (DHS-2) (Nos. 10) Mango (Alphanso)(No. 01) Lime (Kagazi lime)(No. 01) Guava (Lucknow—49)(No. 01) Curryleaf (Suwasini)(Nos. 10)
				Trees:	Teak (Nos. 10)
				Bunds:	Castor (48-1)

Results and Discussion

To analyse the land use pattern of conventional cropping system and after adoption of the Integrated Cropping System, an attempt was made and the results are presented in Table 1. The cropping pattern for the sample farmer with his CCS indicated more than 95 per cent of the total area was utilised for the cultivation of food grains (green gram, maize, sorghum, etc.) and commercial crops (cotton). On the contrary, after

Table 2: Production and Economics of the Conventional Cropping System with Integrated Farming System

Particulars	Conventional Cropping System (2003-04)				After adoption of Integrated Farming System (2004-05)			
	Qty. (q/L)	Total Returns (Rs.)	Cost of Cultivation (Rs.)	Net returns (Rs.)	Qty. (q/L/ Nos.)	Total Returns (Rs.)	Cost of Cultivation (Rs.)	Net returns (Rs.)
Food grains	28.00	16400	5200	11200	15.60	33.650	15800	17850
Commercial crops	2.00	3200	1900	1300	4.00	92000	42000	5000
Oilseeds	–	–	–	–	4.00	65.50	2600	3950
Vegetables	–	–	–	–	91.00	53100	17800	35300
Fruit crops*	–	–	–	–	*	*	*	*
Dairy (milk yield in L)	280 L	2240	1200	1140	420 L	3360	1800	1560
Poultry (Giriraja-20:4) Sale of eggs/aged birds	–	–	–	–	200 eggs+30 kg meat	4400	1250	3150
Production of vermi-compost	–	–	–	–	80	3200	800	2400
Gross total	30.00+ 280 L	21840	8300	13640	114.60 q +420 L+ 200 eggs+ 30 kg meat	113460	44250	69210
Percentage increase over benchmark	–	–	–	–	282 q +50 L +200 eggs+ 30 kg meat	419.51	433.13	407.40
Additional income over benchmark	–	–	–	–	–	–	–	55570
Cost benefit ratio	–	–	–	0.61	–	–	–	1.56

*After pre-bearing period, the returns will increase accordingly.

adoption of IFS with dairy and poultry components, food grains occupied 50 per cent of the total cultivated area followed by the oilseeds and vegetables.

Returns realised and cost incurred by the farmer CCS and IFS were analysed and results are presented in Table 2. The total returns with CCS was Rs. 21,840, wherein farmer had made expenditure of Rs. 8,300 towards the purchase of inputs.

With CCS, farmer had realised net income of Rs. 13,640. Similarly results also inferred that the total returns after adoption of IFS was Rs. 1,13,460 with an expenditure of Rs. 44,250 which has resulted in net profit of Rs. 69,210. From the Table 2, it can be observed that after adoption of IFS module, though the cost of cultivation was increased, correspondingly net profit also increased. The overall net profit earned over the CCS was 407.40 per cent. Further, the B:C ratio of CCS was observed to be 0.61, while after adoption of IFS module it raised to 1.56. Similar benefits of integration of different components in crop husbandry was reported by Sharma and Bhattacharyya (1988).

Conclusion

The integration of different components in crop husbandry favoured physical and economic benefits over the CCS. The risks and uncertainties encountered due to the biotic and abiotic factors have been reduced and module turned into sustainable and reliable module for the effective use of available resources at farm level. This also yielded better returns to the investment made.

REFERENCES

Anonymous (2005). Potential Linked Credit Plan 2004-05 of Haveri district. Published by NABARD, Regional Office, Bangalore.

Sharma, K. and Bhattacharyya, N.K. (1988). Integrating agro-forestry with goat husbandry for the benefit of Rural poor. *Wasteland News*, IV (1): 13-17.

26

Impact of Wasteland Development Programmes in India—An Appraisal on Commissions of Omissions

C. Cinthia Fernandaz and *K. Chandrakandan*

Introduction

The focus on poverty reduction of the millennium and the development goals have given new urgency to agricultural policy. For many years governments have debated the merits of investing in low productive areas, i.e. wastelands. About ten years ago research by the Drylands Research Organization showed that dryland farmers can and do invest in sustainable natural resource management and livehood diversification, under the right conditions, there can be positive links between population growth, agricultural intensification and the improvement of livelihoods.

It has been estimated that about two-thirds of the cultivable area in our country remain as drylands and about 20 per cent still left as wastelands, due to obvious reasons.

Mishra (2005) has overviewed the wasteland mapping for sustainable development. The semi-arid tropics include parts of 48 countries in the developing world. Most of India come under it. They are characterised by unpredictable weather, long dry seasons, inconsistent rainfall and soils that are poor in nutrient. Hence, for optimum use of land resources, wastelands need to be mapped and brought under productive use.

Development of wastelands is an important programme which includes among other things, checking land degradation, increasing bio-mass availability and putting wastelands to sustainable use.

Development Programme for Wastelands

Sustainable use of these wastelands can be made possible through various projects and allied farming activities. Due to meagre rainfall conditions cattle and animal rearing is not possible. Government has initiated the wasteland development.

Integrated wasteland development programme (IWDP)

This scheme has been under implementation since 1989-90 and has its effect with National Wastelands Development Board in July 1992. From 1st April 1995, the scheme is being implemented on watershed basis under the common guidelines for watershed development. The scheme also helps in generation of employment in rural areas besides enhancing people's participation in the wastelands development programmes at all stages. It consists of 247 IWDP projects in 25 states in a total outlay of Rs. 778.12 crore to treat total project area of 15.98 lakh hectare at various stages of implementation. Out of these the projects taken up before 31.03.95 have more or less been completed. During the year 1999-2000, 38 new watershed projects for development of 3.71 lakh ha costing Rs. 148.46 crore, over a period of five years have been sanctioned. There is a provision for a greater role of Panchayat Raj Institution, Self Help Groups and User Groups. IWDP is a 100 per cent central sector scheme. It is proposed to change funding pattern from 100 per cent central funding on sharing bases in the ratio of 75:25 to ensure greater involvement of state government.

The cost norm at present is Rs. 4,000 per ha for a project under this scheme. This leads to equitable sharing of benefits and sustainable development. The major activities are:

1. Soil and moisture conservation measures like terracing, bunding, trenching, vegetative barriers, etc.
2. Planting and sowing of multi-purpose trees, shrubs, grasses, legumes and pasture land development.
3. Encouraging natural regeneration.
4. Promotion of agro-forestry and horticulture.
5. Wood substitution and fuelwood conservation measures.

6. Measures needed to disseminate technology.
7. Training, Extension and Creation of greater degrees of awareness among the participants.
8. Encouraging people's participation.

Drought prone area programme (DPAP)

The Drought Prone Area Programme (DPAP) is one of the areas development programmes launched by the government in 1973-74 to tackle the special problems faced by those fragile areas which are constantly affected by committee (1994), the programme is being implemented on watershed basis. The responsibility of planning, executing and maintaining the watershed projects is entrusted to local people's organisation specially instituted for the purpose. In view of the large problem area to be treated and the present level of financial allocation, the programme has to be taken up on a continuing basis for several years. Presently, 947 blocks of 155 districts in 13 states are covered under the programme. The states are Andhra Pradesh, Bihar, Gujarat, Himachal Pradesh, Jammu and Kashmir, Karnataka, Madhya Pradesh, Maharashtra, Orissa, Tamil Nadu, Rajasthan, Uttar Pradesh and West Bengal. Currently, under DPAP, 6515 watershed projects have been targeted for development over a 4-5 year period against which 6002 projects, covering an approximate area of 30 lakh hectares are under various stages of implementation. The project outlay for there 6002 projects of Rs. 1191 crore with an average project size of 500 ha and average cost of Rs. 20 lakh. The expenditure on DPAP was being shared equally by the Central and State governments on 50:50 basis till 31.3.99. This funding pattern has been revised to 75:25 w.e.f. 1.4.99.

Desert development programme (DDP)

Over the years, the increase in human and livestock, population in drought prone and desert areas has placed. The natural resources of the regions are under great stress. The major problems are continuous depletion of vegetative cover, increase in soil erosion and fall in groundwater table. On the recommendations of the National Commission on Agriculture in its interim report (1974) the Desert Development Programme (DDP) was started in the year 1977-78. The programme was started both in the hot desert areas of Rajasthan, Gujarat and Haryana and the cold desert areas of Jammu and Kashmir and Himachal Pradesh from 1995-96, the coverage has been extended to few more districts in Andhra Pradesh and Karnataka. Since inception till 1994-95 over 5.5 lakh hectares of area was treated under this programme, presently 227 blocks of 36 districts in 7 states are covered under the programme. Currently, under DDP 2202 Watershed Projects covering an approximate area of 11 lakh hectares have been targeted for development over a 4-5 year period. 390 special projects with a total outlay of Rs. 97.5 crore for 10 districts of Rajasthan for sand dune stabilisation, shelter belt plantations, etc. have been sanctioned and an amount of Rs. 10.97 crore has been released. In Sandy areas, DDP was 100 per cent centrally sponsored, whereas in non-sandy areas, the expenditure was being shared by the Central and State governments in the ratio of 75:25. This funding pattern has been revised to 75:25 from 1.4.99 uniformly in all areas.

Investment promotion scheme (IPS)

The scheme was launched in 1994-95 in order to stimulate involvement of corporate sectors and financial institutions for the flow of funds for development of non-forest wastelands. Under this scheme, central promotional grant for general category is limited to Rs. 25 lakh or 25 per cent of the project cost for on-farm development activities whichever is less subject to the condition that the promoter's contribution in the project shall not be less than 25 per cent of the project cost. In the case of small farmers it is 30 per cent and in case of marginal farmers and SC/ST it is 50 per cent.

Technology development, extension and training scheme (TDET)

Technology support is extremely vital for the success of a land based programme, especially in the development of wastelands. The scheme is being implemented through ICAR, State Agricultural Universities, DRDAs and government institutions having adequate institutional framework and organisational back-up. Successful implementation of this scheme is expected to bridge the gap between the existing technologies and the need relevant to the latest situation. Under this scheme 100 per cent central grant in admissible to implement projects on wastelands owned by govt. public sector undertakings include Universities, Panchayats, etc. In the case of projects on wastelands of private farmers corporate bodies, the cost of the project requires to be shared on the basis of 60:40 between DOLR and the beneficiaries.

Supports to NGOs/VAS (Grant-in-aid) Scheme (NGOs)

The objective of the scheme is to create awareness, application of appropriate technologies for the development of wastelands and training for increasing capability building. The extension and publicity are other important components of the scheme. Under this scheme, 100 per cent grant is provided to registered voluntary agencies, cooperatives, Mahila mandals, Yava mandals and other similar organisations for undertaking work directly or indirectly encouraging afforestation and wastelands development. New projects under the scheme are now being sanctioned by CAPART. The department has transferred all fresh applications and ongoing projects under the scheme to CAPART except 75 cases which are being implemented by cooperatives and trusts, etc.

Watershed development projects

As per old guidelines of Employment Assurance scheme 50 per cent of the funds released under DPAP and DDP and up to 40 per cent of the funds released under EAS were supposed to be spent on watershed development projects and other related activities. The various states were following it and a certain percentage of the funds was being spent of watershed development projects. The states have in the meanwhile, created a liability for completion of watershed development projects under EAS. As per latest figures the total area taken up under EAS prior to 31.3.99 for completion of watershed development projects comes to 63.50 lakh ha for which central share funds will be released by the DOLR.

Training for watershed development

The main thrust under Common Guidelines for watershed development for the major schemes of IWDP, DPAP and DDP is on people's participation by involving them in the decision making. Implementation and monitoring of the projects through formal and informal groups and village level institutions. The project estimate provides for 5 per cent for providing training to the functionaries for successful implementation of the projects. For successful implementation of these Area Development Projects, government has constituted a committee on 'Training for Watershed Development' under the chairmanship of Shri V.B. Eswaran. The Eswaran Committee emphasised the need for massive training programme for different level of functionaries in the implementation of the projects. Accordingly, guidelines have since been formulated to provide financial assistance to the SIRDS, ETCs and other Training Institutions, identified by the state governments for impacting training on watershed management.

Studies on watershed development

Ramanna (1999) conducted a participation study on watershed development in the southern dry region of Karnataka state. It was a comparative study of watersheds of government vis-à-vis NGO. He has selected one good functioning watershed and a bad one in the NGO and the same one in the government sector, 50 respondents in each one so totally 200 respondents. He has mentioned the constraints faced by the beneficiaries.

Constraints faced by the farmers of IWDP

It could be observed that nearly two-thirds (62%) of the government watershed beneficiaries have expressed lack of knowledge as the main constraint in participation followed by uneven distribution of incentives (58%), supply of poor quality inputs and material (52%) and groupism and politics at village level (48%). Almost an equal percentage of beneficiaries indicated lack of initiative and motivation (44%) and poor quality of work (42%), by implementing agency as the reasons. Exactly one-third 33% and 21% of the beneficiaries have felt that lack of motivation from the village leaders and improper locations of soil and water conserving structures as the constraints in participation.

With respect to NGO and watershed, one-third of farmers expressed lack of knowledge (33%), followed by poor quality work (32%) and supply of poor quality of inputs and materials (30%) as major constraints in participation.

Constraints in Participation

Sl. No.	Problems	GO			NGO		
		Frequency	%	Rank	Frequency	%	Rank
1.	Lack of knowledge about the programme	62	62.0	I	33	33.0	I
2.	Uneven distribution of incentives	58	58.0	II	22	22.0	IV
3.	Supply of poor quality material and inputs	52	52.0	III	30	30.0	III
4.	Groupism and politics in the village	48	48.0	IV	18	18.0	IV
5.	Lack of initiative and motivation	44	44.0	V	20	20.0	V
6.	Poor quality of work	42	–	VI	32	–	II
7.	Lack of motivation from village leaders	33	–	VII	10	–	VII
8.	Improper location of soil and water conservation structure	21	–	VIII	8	–	VIII
9.	Planning is not concerned to local needs	13	–	IX	5	–	X
10.	Lack of time	8	–	X	–	–	–

Suggestions from the beneficiaries to enhance participation in IWDP

Nearly 68 per cent government watershed respondents suggested conducting more educational activities followed by involvement of villagers in planning (56%) and equitable sharing of benefits among the beneficiaries (55%) for effective participation. Forty eight per cent and 46 per cent wanted to give equal importance to all members and to increase the frequency of visits of field staff respectively. Almost on equal percentage of beneficiaries opined to creation of awareness and interest (42%) and provide inputs at free of cost (41%).

Viable approach towards the development of wastelands

- Assign suitable land for each project in different parts of the state.
- Lease 50 per cent of the required land to corporates willing to invest in the approved agro-enterprises.
- Assign the balance 50 per cent to rural families for cultivation as a registered crop and supply of produce to the power plants.
- Cooperative power plants can also be established in regions such as Maharashtra where agricultural cooperatives have proven effective.
- Nationalised banks can be directed to extend crop loans to farmers based on registration of their crops with the local power plant, in the same manner as crop loans are now commonly extended to sugarcane cultivators whose crops are registered with nearby local sugar factory.

Extension methods for appropriate usage

Group contact and mass contact methods are more efficient towards the rural people in creating awareness among the people skill and method demonstration are more useful as far as the development of skill is concerned. Farmers participated in IWDP watershed programme and were convinced about the dryland technologies and started practicing them. They have doubled the yield of their land after participating in IWDP. They became watershed promoter of their area in 1997-98. They concentrated on non-arable land also. They planted mango, sapota, guava and forest species of teak, tamarind, jack and jambu. Their cropping intensity turned to 170 per cent and their benefit: cost ratio was 1.90 in 1997-98.

Conclusion

Research into policy

The focus on poverty reduction of the Millennium Development Goals has given new urgency to agricultural policy. Policymakers must be convinced that dryland agriculture and ecosystems are worthy of investment mainly involving the equal people.

Interactive meetings should be initiated in each country from village through local government to national level by providing a forum for impartial debate conducted in local languages, pointed towards priorities which reflect the interests of local people. Some key priorities are:

- Improved access to markets is needed if dryland farmers are to make the most of new opportunities. This involves improved access to infrastructure, institutions and information.
- Dryland farmers can only take advantage of growing urban food markets if national currency management and agricultural subsidy policies are appropriate.
- Economic policies affect the ability of farm household to accumulate wealth and the direction of their investments. Policymakers need to consider these limitations when forming economic policies.
- Targeting policymakers should take account of the level at which research policy influence is desired. international, national, sectoral and Sub-sectoral levels are required at different approaches. A bureaucratic approach is appropriate for linking farmers with Professionals. Relationship between researchers and farmers need to be flexible to different conditions. Local endorsements, as well as the quality, are very important if research is to influence policy. This is very different from the normal expectation that villagers should accept the views of experts without question. Engagement with participatory processes is an essential activity for researchers who want to strengthen the usefulness and effectiveness of policies.

REFERENCES

Mishra (2005). Technology Development, Extension and Training Scheme, Google search.

Ramanna, K.N. (1999). People's participation.

Socio-Economic Concerns in Wasteland Management

27

G.P. Reddy

ABSTRACT

In this chapter emphasis are placed on problem-oriented surveys, planning, social-economic concerns and implementation of wastelands. Various challenges and constraints in developing economies have been discussed and suitable measures have been suggested. The management possibilities and alternatives to be considered at the planning stage are briefly spelled out. Planning mechanism should have a joint entity with decentralised implementation. This chapter also explains the needs of establishing database, monitoring and evaluation system in addition to data collection.

Wasteland management planning in developing countries is usually hampered by limited manpower resources and time. Therefore, the work should be carried out as practical and economic as possible. The final products should be oriented to problems and their management rather than theories and academic studies.

As wasteland problems grow with population and time their management tasks are almost endless. Planners must keep in mind that planning is a continuous process and there is no single plan or programme, regardless how sound it is that can solve the problems forever and completely. When major problems are tackled and databases are established, such plans should be considered useful and beneficial. Too ambitious a plan causes non-approval and non-operation. Also, a small success is better than a big failure. Government goals may not always coincide with the interests of farmers or wasteland inhabitants.

Challenges to Wasteland Planners

To start any wasteland planning, planners should be prepared themselves to face many challenges. They need also to determine what principles should be best observed and what courses taken. The following examples include some major challenges frequently confronting the planners in the developing countries:

- Political views may differ considerably from those of technical persons on types of management work, priority areas and timing.
- An ideal physical plan or a most effective work may not always be socially receptive by local communities or farmers.
- Planning from bottom up may cost lots of time and energy in wastelands where populated with illiterate farmers, yet, involvement of local communities in planning process in a necessity.

There are no universal answers to these challenges. Each of them should be carefully examined in the context of the given conditions and the existing environment. As stated in a World Bank paper, wasteland project design is inevitably a compromise. There is a need to select a practical approach to obtain early agreement and initiate operation for benefiting both the people.

Quick Identification of Problem Areas

National reconnaissance and watershed classification

At the national level, a quick reconnaissance type of survey, assisted by remote sensing techniques, is often sufficient for identifying major wasteland problems and areas. Using existing date and knowledge can also make a quick classification of a nation's wastelands. They can be classified into i.e. reservoir wastelands, forestry and wild land wastelands, agricultural wastelands, municipal wastelands, etc. for ready pinpointing of their major problems and for establishing management policies and priorities.

Problem oriented surveys

At the local or wasteland level, further surveys or investigations are needed to obtain basic information for formulation management plans. These surveys should be problem-oriented, in a way similar to a doctor's approach to a patient. Healthy patients or wastelands should not be ignored but they should be put on routine care or maintenance while special attention or urgent treatment is given to serious ones. The "right medicine for the right illness" will not only save money and time but also lives.

Wasteland problems can generally be grouped into several categories as follows:

- Physiographical problems (e.g. steep slopes, heavy rains, excessive run-off, problem of soil, etc.)
- Resources use problems (e.g. shifting cultivation, forest destruction, fire, over-grazing, uncontrolled mining, poor road construction, etc.)
- End problems (e.g. erosion, sedimentation, flood, water pollution, water shortage, etc.)
- Socio-economic and other problems (e.g. illiteracy, low acceptance of innovation, labour shortage, land tenure, poor infrastructures, etc.)

Detailed surveys or investigations should be centred on the major wasteland problems identified during preparatory missions or preliminary investigations.

Considering management possibilities

A practical wasteland plan should consider, at the very beginning, the management possibilities and suitable strategies. For example, some foreign experts of outsiders, when faced with local wasteland problems, may simply suggest moving the people out of the wastelands. It may sound attractive but they don't realise the difficulties and expenses of resettlement, not to mention creating social problems. Solving a problem in one place by creating more problems at other places is not a wise policy to pursue.

Management possibilities, which need to be taken into good considerations, are suggested as follows:

- *Technological aspects*: Whether the present technology is sufficient for coping with major problems in the wastelands or more local trials are still needed. What will be the effect of the transfer of exotic technology, etc.
- *Available resources*: What are the possible sources for obtaining resources, monetary and the like, for the proposed wasteland programme? The magnitude of the programme should not exceed the expected resources.
- *Institutional ability*: The strengthening of any present institution is quite limited because it needs time to train technical staff. The programme should grow only as fast as trained and experienced staff are available.
- *People's acceptance*: Although this will depend on a continuous effort of education and extension, yet extreme care should be given to those management practices which are against local culture, religion or tradition. People's acceptance is a key to the success of most wasteland programmes, but it is always a slow process.
- *Expected outputs*: What really can be accomplished at the end is often an ambiguous and neglected subject. Exaggeration of results may attract funding but it eventually will cause despair, to many funding agencies. The planner should be frank at the planning stage and realise that there are limits of management practices.

Wasteland management and people

The focal point of any program is the betterment of the people. Thus, the wider socio-economic concerns of any package of technology ought to include the size of population, population of ethnic groups or communities, composition and structure of households and basic and secondary needs of the people. High population growth and the failure of the governmental programs to keep pace with population growth land to such problem as unemployment, underemployment, inequalities in income and assets and over-exploitation of natural resources. Poor people have to live more on natural resources and they get caught up in a vicious cycle of removing more from fewer resources. The best example is Jhum cultivation. At present no in-depth survey in this regard is being done before planning a project, which could assist in locating persons also might act as catalyst in the program and encourage people to participate.

Land capability and social needs

Improved land use is not an independent to be pursued relentlessly regardless of all other factors. Worldwide, there is a starting correlation between poverty and abusive land use. But the poor are not always to blame. Their immediate problems are often too pressing. Therefore, to solve the gigantic problem of soil and water conservation, it is imperative to match the land's needs with the needs of the landholder. For this to occur, an in-depth survey of available natural resources as well as socio-economic factors in needed.

More emphasis must be put on the tenure system, leasing system, insecurity and capability of resources. Policies and programs should answer such questions as: which crops will grow in which areas? Where will new arable land be developed? What agro-forestry practices, if any, are feasible? How the cost of protection be saved, etc. It is not enough to impose these changes on the people who practice them.

Planning Mechanisms

Extent of coordination

The extent of coordination among various agencies depends on the degree of integration of a wasteland programme. Generally speaking, because of the composite nature of wastelands, no single agency could plan or manage a wasteland fully. Coordination is almost a must. Nevertheless, the word "coordination" is more often spoken than acted upon in the developing countries due to lack of qualified staff and resources in each agency. Only when there are surplus staff and its main task is being taken care of will an institution render assistance to the others. Usually, the least efficient agency limits the whole progress of work.

Involvement of local communities

As mentioned before, this is a necessary but difficult task. If there are local bodies such as local governments, farmers' associations, private interest groups and specially organised districts, their appropriate representatives can be involved in wasteland planning. Care should be given not to involve those who are too strongly into party politics that to involve them would cause trouble or make others non-cooperative.

Another technique is to directly poll farmers or inhabitants by specially designed surveys. Their answers may not be very useful unless the questionnaires are properly designed and tested.

Organising proper bodies for planning

First, there should be one agency designated to initiate the work. Usually this agency should have its chief responsibility in managing wastelands or have its own interest in wasteland products (i.e. irrigation, electricity authority or municipal governments). A steering committee should be organised involving representatives of essential Agencies and local communities with the organiser as the convener. The committee thereafter monitors the processes of design, survey, analysis and final reporting. If required, the committee will be involved in implementation and evaluation.

Below the committee, a number of field survey teams are usually needed. A team can have members from different agencies but its leader should be drawn from an agency related to the survey subject. For instance, the leader of a soil conservation agency although members may come from extension, crops and soil department and farmer's association, etc. The leader will not only report their progress to the steering committee but also serve as a bridge from his mother organisation to the committee.

Periodically, all the teams need to meet and report their progress and problems to the steering committee. Any inter-team matters should also be settled in the committee meetings. Funds for planning can be provided mainly by the chief organising agency because they have more interest than others or it may be shared among coordinated agencies. At the end of the planning, each agency should agree to an overall plan consisting of work schedule, staff needs and budget. When no external aid is available each should share the cost and programme in their regular budget year by year, according to the overall plan. The programme will eventually be carried out by each responsible agency in the wasteland with mutually understood objectives and overall supervision. Sometimes, a liaison office is needed in the field to represent the steering committee.

This kind of joint and coordinated planning and decentralised implementation proves much more effective in many developing countries. It is better than piecemeal approaches and staff-borrowing type of undertakings. As long as each agency considers it is its own responsibility and is not merely to do others favour, the work will be done more efficiently and fruitfully.

Data requirements and collecting techniques

Data required for wasteland planning vary according to management objectives, wasteland problems and given conditions. Generally, the following categories of information are needed:

- Physiographic data including location, elevation, soils, geology, land forms, slopes drainage patterns, etc.
- Socio-economic data including demography, land tenure, farming systems, education, infrastructures, human resources, farm enterprises, rural employment, production, income, marketing, transportation, credits, labours, etc.
- Institutional and cultural data including policy and administration, extension services, farmer's organisations, community and private groups, traditions, religions, cultural practices, acceptance to innovations group actions, etc.
- Management oriented data including wasteland problems, environment impacts, land management

techniques, treatment needs, unit cost, sectoral cost, cash flows, work schedule, financial arrangements expected benefits and results, etc.

Feasibilities, Alternatives and Recommendations

Assessment of feasibilities

Assessment of project feasibilities by economic means is one of the measures often employed in planning. However, economic viability is not the only criterion for decision making. Wasteland projects, unlike simple engineering works, have their special characteristics which are not easily assessed by monetary means, i.e. long-term and intergeneration benefits, externalities, spatial distribution of costs and benefits intangible and indirect benefits, etc.

Besides the usual cost and benefit analysis with and without project and the calculation of internal rates of returns, etc. the social values or benefits should also be added for overall assessments of wasteland projects. For developing countries, social benefits derived from increased employment, equitable distribution of wealth improving the farming environment and farmer's well-being and providing better infrastructures, etc. should be weighed heavily in the final assessment.

Selection of alternatives and strategies

There are many ways to achieve management goals. During the planning stage, possible alternatives should be explored so that the best alternative can be chosen. For instance, if the main objective is to restore vegetation on the steep slopes for protection, expensive reforestation may not be the best answer. Alternatives could include protection of the existing land for natural regeneration/succession or supplementary planning or seeding, etc. Some technological alternatives need field trials, which can be built into the project at the beginning of the implementation stage. Alternatives in cost and effect, in incentive needs vs. education, in group actions vs. individual assistance, among others, should also be well considered and selected.

For wasteland project to be successful, strategies for effective extension for streamline operations, cost control, inducing farmer's participation and for work maintenance, etc. should be well chosen before initiation of implementation. Many constraints in physical, technical administrative and budgetary aspects should therefore be clearly identified and resolved in the planning stage.

Making realistic recommendations

The wasteland planners, being not always the future project executors, tend to make recommendations good in theories but questionable in practices. Examples are not lacking such as to move large farming communities out of wastelands, to complete treating a huge wasteland in 3 or 4 years or to ask government to increase staff out of proportion to its existing structure, etc.

Planning should by no means be only for planning's sake. To sell the plan to either the government or to any international aid agency, recommendations have to be relevant, responsible and realistic.

Monitoring and evaluation

Monitoring and evaluation are critically important in realizing the objectives of developmental programs and projects, particularly wasteland projects having multiple objectives. In India, integrated rural development projects, including soil and water conservation and wasteland progammes are being monitored at four levels—Central, state, district and block. The Ministry of Rural Development and/or the Ministry of Environment and Science reports from state level. Rural development aggregate or couple this reports. This monitoring is done mainly on a secondary database. In most cases, the question of what should be monitored and at what stage is not clear to the field staff. These reports rarely are of any use in terms of suggesting mid-term corrections. Often, monitoring gets underway after, rather than before a project has begun. Monitoring of socio-cultural factors and of infrastructure becomes increasingly important but neglected either because of rigid, target-oriented progammes are lack of proper guidance. In fact, there are no separate monitoring and evaluation cells at district level. Monitoring of project effects after withdrawal of a project is rarely done, which could be useful for future projects.

Evaluation if a process of determining systematically and objectively the relevance, efficiency, effectiveness and impact of activities in light of their objectives. It is an organisational process for improving activities still in process and for adding management in future planning and decision making. But in country, most of the evaluation studies are confined to achievement or fulfillment of targets and efficiency of implementation. Evaluation of these programmes must include the process of percolation spill over effects, lacunae in the program and the phenomenon of non-participation on the part of local people in the program. Lack of standard or well-developed evaluation methodologies for evaluation of wasteland programmes clearly has hampered progress. This indicates that there should be a team of social scientists at least at the district level that can develop a suitable methodology for evaluation of programs and update data banks at the district level. Involvement of social scientist at all stages of a project or program will not only increase program efficiency but also indicate needed improvements.

Conclusion

Wasteland management technologies must be compatible with socio-economic concerns in a developing society. Efficient and properly linking technology and various socio-economic aspects is critically important. Wasteland technologies will gain acceptance among the rural masses only when these socio-economic considerations are adequately addressed. A strong data base is needed at least at the district level as are effective coordination between various governmental and non-governmental agencies; other establishment of a planning, monitoring and evaluation cell at the district level; surveys and their documentation before planning; development of alternative set of technologies and the active involvement of local people in preparation execution and maintenance of structures in a wasteland. Otherwise, huge amounts of scarce resources could be wasted on ill-conceived, poorly designed, and badly executed wasteland programs.

REFERENCES

Gregersen, H.M. and Books K. (1978). "Economic Analysis of Watershed Projects." Paper presented to 12th session of the working party on the management of mountain watersheds, Rome.

Food and Agriculture Organization, United Nations (1986). *Guidelines for Watershed Management*, Rome, Italy.

Sheng, T.C. (1981). Physical planning for watersheds FAO/UNDP: Project working paper 17.

Srivatava, R.C., Sahoo, N. and Reddy, G.P. (2001). *Training manual for watershed management technology for Eastern region*, Water Technology Centre for Eastern Region, Bhubaneswar.

Traditional Dry Land Farming Practices of Farmers in Tamil Nadu

28

K. Thangaraja, C. Karthikeyan J. Venkatapirabu and *R. Rajasekaran*

Introduction

Dry land agriculture in India is always a challenge, since crop production in dry land depends on the vagaries of monsoon. Dry land areas receive less than 750 mm rainfall. Out of the net sown area of 136.18m ha, the dry land accounts for 93.13 m ha (68.4%) and contribute 40 per cent of total human population, 60 per cent of cattle heads and 44 per cent to total food production in India (Kannaiyan et al., 2001). In Tamil Nadu, area under dry farming constitutes 52 per cent of the total cultivable area and contributing to 40 per cent of total food production. The productivity of crops grown in dry land is not only low but also remains stagnant over years. Most of the areas of Tamil Nadu come under semi arid tropical climate, except the hilly regions and East Coast. Out of 5.50 m ha of net sown area, nearly 3.20 m ha are rain fed (Kannaiyan et al., 2001).

The term Indigenous knowledge is used synonymously with 'traditional' and 'local knowledge' to differentiate the knowledge developed by a given community from the international knowledge system, some times also called as 'western system' generated through universities, government research centers and private industries (Somasundaram 1995). The decrease in cultivable land was due to increased population and their requirements for living space. The achievement was because of high yielding varieties, which were highly input responsive. Farmers used chemical and synthetics pesticides, fertilizers and growth promoters in high quantities with a motive to get high yields. In this process, the soil and environment were got polluted due to injudicious land utilization with high intensity cropping and dumping of chemical and synthetic pesticides, fertilizers and growth promoters.

At this juncture, there was an additional responsibility on agricultural research system to minimize the soil and environment pollution along with keeping intact of the primary objective of achieving the targeted food grain production. To satisfy these two requirements, the agricultural system aimed at eco-friendly low-cost farming methods in integration with modern technologies (Lakshmana, 2000).

Traditional practices are relevant under the changing scenario in rainfed agriculture and also to impress upon the need for blending the traditional practices of risk management with modern practices of high production. This knowledge in today's parlance is called local knowledge/traditional knowledge or simple indigenous knowledge (Balasubramanian, 1992). The interest in traditional knowledge is growing with considerable momentum. So, also in the case of dry land agriculture where the modern knowledge alone is being considered inadequate to overcome the problems. There is undoubtedly a need to initiate systematic efforts for collection of the traditional practice from different areas. So that these could be catalogued for wider use. The ITKs would be helpful to develop ecologically compatible and socially accepted technologies in different areas of crop and animal culture. Keeping these facts in view, documentation was made on the indigenous dry farming technologies that were adopted traditionally over years by dry land farmers.

Materials and Methods

The study was conducted in Dindigul district of Tamil Nadu. Reddiarchathiram block was purposively selected for the study, since the main occupation of most of the people was agriculture, which depended on poor and erratic rainfall. Majority of the farmers in the sampled block had been practicing dry farming. From this block, three revenue villages were selected randomly. The sample size was fixed as 90 farmers consisting 45 Maize growers and 45 Sorghum growers. The selection of 30 farmers was done at random in each of the selected revenue villages.

Thus, 10 farmers were selected from each of the nine villages, covering three revenue villages using simple random sampling technique. Participatory Rural Appraisal (PRA) technique was used to document the indigenous practices adopted by dry land farmers in these selected villages. Key informants, farmwomen, youth, labourers formed the sample for data collection in each of these selected villages.

Result and Discussion

Indigenous practices followed by dry land farmers

The dry land farmers practiced many low cost but effective technologies, which were adopted over many years in their locality. They perceived that these indigenous practices are effective and influenced the cost of cultivation as well as the crop productivity. The following are the indigenous practices followed by the Sorghum/Maize farmers in the study locale. Most of these ITKs are practiced by small/marginal farmers who mostly depended on local resources for raising crops. However, these ITKs were not commonly adopted by big farmers who mostly depended on scientific modern practices only. The documented Indigenous practices were classified under 6 major head as given below.

A. Preparatory cultivation practices

1. Summer ploughing

Majority of the Maize and Sorghum growers had adopted summer ploughing 2-4 times over years. Ploughing the land soon after the receipt of down pour. The advantages of these practices are,

- (i) Storing the rain water i.e. moisture retension in ploughed field for longer period.
- (ii) Destroying the eggs, larvae and pupae of pest in soil.
- (iii) Eradication of weeds.
- (iv) Control the soil erosion
- (v) Reduction in the number of tillage operations during the time of sowing.

B. Seeds and sowing

2. Maize seeds soaked in cool water

Before sowing, the Maize seeds are soaked in cool water for 2-3 hours. It could initiate early germination and higher germination percentage.

3. Soaking Sorghum seeds in boiled cool water

Sorghum seeds are soaked in Luke warm water for 8 hrs (or) over night before sowing. The purpose is to improve the germination efficiency.

4. Soaking Sorghum in cow urine

The Sorghum seeds are soaked in cow urine over night before sowing. The purpose of this practice is to induce drought tolerance in seeds.

5. Cow dung coating for cotton seeds

Cotton seeds are coated with cow dung before dibbling. This helps in intacting the fuzz in the seeds and set ease to dibble the seeds.

6. Cotton seed coated with red soil

Cotton seeds are coated with red soil and then sun dried before sowing. Advantage of this practice is to improve the germination percentage and becomes easy in dibbling the seeds.

7. Cotton seeds mixed with rice poridge and cow dung

Cottonseed mixed with required quantity of rice poridge with cow dung before sowing. An hour of shade drying is done for the purpose of separating the seeds during sowing period.

8. Gourd seeds mixed with cow dung

Pumpkin and bitter gourd seeds are mixed with cow dung before sowing of the seeds. Purpose of this is to improve the seed germination and induce the drought tolerance.

9. Removal of fuzz from cotton seeds

Cotton seeds weighing five kgs were mixed with ¼ lit. Butter milk and cow dung (½ kg). The purpose is to remove the fuzz from cotton seeds and also to improve the germination.

C. Manure and manuring

10. Sheep penning

This is the practice of keeping the sheep penned overnight on the cultivated land. So the dung and urine excreted could be directly absorbed by the soil. It is followed in summer when the land is free from crops. Payment of 50 paise / goat / day has to be made by the beneficiary to the goat rearer. One acre of land can be covered for penning by 100 goats for a period of 40 days.

D. Mixed cropping

11. Sorghum and Maize raised as a mixed crop with lablab

The lablab is mixed with broadcasted Sorghum and Maize crops. The purpose is to get additional yield and improve soil fertility through nitrogen fixation.

12. Raising Agathi crop as a border crop in Sorghum field

Agathi (*Sesbania grandiflora*) has been used as a fodder crop. So, it is raised as a border crop in Sorghum field.

E. Plant protection measure practices

13. Beating empty drums to ward away birds

This is the practice of blaring empty drums in Sorghum and Maize fields by engaging a man for this purpose. This practice scares away the birds from the field which would otherwise damage the crops.

14. Bird watch

Polythene sheet is tied to a long pole and placed in the center of the Maize field. During windy occasions, polythene sheets flutter and produce sound to ward off birds. Noise from a tape recorder and cassette strips are also used for scaring of birds in Maize and Sorghum fields in dry land area.

15. Throwing stones with the help of leather rope

Stones are kept in a leather pouch tied with strings. Catching on to one end of the string swirls this around. On release of the string the stones will be thrown away at a greater distance (20 m), which would chase the birds away from the field.

16. Palmyra leaf is also used to produce blaring sound to alarm birds from the Sorghum and Maize fields.

17. Raising castor crops as a border crop in rainfed cotton field

The castor is grown as bund crop in cotton field. It also acted as a trap crop. Because, the castor crop readily traps pests like *Spodoptera litura.*

18. Cotton inter cropped with red gram

Red gram is intercropped with rainfed cotton i.e. red gram was sown in the bund. The purpose is to reduce the pests like bollworms, hoppers etc.

19. Neem seed kernel is soaked in water kept over night. Next day morning, the suspension is filtered and sprayed by adding sufficient quantity of water. It controls the pest like Hopper, Jassids etc in cotton crop.

F. Indigenous post harvest practices

1. Coating pulses with red soil: Before storing for latter use, pulses such as mochai, red gram, black gram and green gram seeds were coated with red soil and the main purpose of this practice is to prevent the pest attack and easy separation of kernels and also to increase the storability of seed for longer period.
2. When the Sorghum seeds are stored for seed purpose, the seeds are mixed with sand. The purpose is to store for longer period by protecting it from storage pest.
3. For seed purpose Maize, Sorghum was harvested separately. Ear heads were threshed and stored in gunny bags along with the husk to protect from the attack by storage pests.
4. Cow dung ash is mixed with the Maize seeds during storage. Purpose is to control the storage pests and to increase storability, without damage.
5. Storing the grains in Mudkudhir: This is the practice of storing the grains (Sorghum, Maize) in "mudkudhir", a container made of mud. The purpose is to store the grains for longer period preventing storage pest attack.

6. Pulses (Bengal Gram, Green gram, Black gram) seeds are soaked in water for over night and sun dried. Purpose is to break the seed for consumption purpose and also increase the storage life.
7. Black gram seeds mixed with castor oil helps in removal of seed coat from the seed.
8. Seeds of Sorghum, Maize, Mochai, Cowpea, Black gram, Red gram, Green gram, etc. were mixed with 10% BHC powder before storing. Purpose is to prevent it from storage pest.
9. Cereals like Sorghum, Maize are well dried and stored in storage bin. With an interval of 30 days, sun drying is done. Purpose is to remove the moisture from the seeds and prevent attack from the storage pest.
10. Sorghum Maize stalks after harvest, are cut and then tied at top portion and sun dried for more than 15 days in the field itself. Dried stalks are stacked and arranged layer-by-layer. Top portion is covered with paddy straw and its purpose is that it is used as feed for animals (dairy, bullock, buffaloes). In this way, bye-products are stored for longer period without damage.

Perceived benefits due to adoption of ITKs by Farmers

The farmers were asked about the perceived benefits felt by them as a result of the adoption of Indigenous dry land practices. The results are presented in the Table.

Perceived Advantages of ITKs by Dry Land Farmers

S.No	Advantages	%
1.	Eco friendly in nature	60
2.	Low cost technology/Economic advantage	60
3.	Easy and well known to adopt the practices	55
4.	Effective utilization of locally available natural resources	50
5.	Labour and time saving technologies	52
6.	More compatible to local situation	43

It was understood that benefits due to adoption of ITKs by farmers in dry land area.

Conculsion

Indigenous practices that are adopted by farmers in dry land farming were documented. These can also be used in future by other dry land farmers else where after proper and scientific validation. Even though there are lots of bottlenecks in the use of traditional practices, they can be solved by proper validation. Thus, dry land farming can be sustained by combining modern scientific practices along with the traditional indigenous practices.

REFERENCES

Balasubramanian, P. (1992). "Indigenous knowledge use in dry lands—An exploratory study", Unpub. M.Sc.(Ag.) Thesis, TNAU, Coimbatore.

Somasundaram, S. (1995). "Indigenous knowledge in farming system", Unpub. Ph.D. Thesis, TNAU, Coimbatore.

Lakshmana, K. (2000). "Documentation and analysis of Indigenous farm practices on major crops. An inventory." Unpub.M.Sc.(Ag.) Thesis, TNAU, Coimbatore.

Kannaiyan, S., Thiyagarajan T.M., Subramanian M., Balasubramanian T.N. and Selvaraj R.. 2001. *Dry land green revolution in Tamil Nadu: The Perspectives*. Coimbatore: Tamil Nadu Agricultural University Press, pp.10-12.

Rainfed Agricultural Development in Tamil Nadu

29

R. Balasubramanian and *N. Venkatesa Palanichamy*

Introduction

Rainfed agriculture plays a crucial role in the agricultural economy of India and Tamil Nadu. As green revolution has largely by-passed rainfed crops, several attempts were made in the past to increase the productivity and profitability of rainfed crops. A major challenge in rainfed areas is not only the low average productivity of crops but also their vulnerability to high risks. In the absence of technologies that will ensure successful cultivation with extremely limited moisture availability and appropriate institutional arrangements to insure the risks associated with rainfed production, agricultural development in rainfed areas remains a challenge. In view of the stagnation in the growth rate of yield of irrigated crops in recent times, improving the productivity of rainfed crops is the only option to sustain the overall agricultural growth. This mission document on agricultural development in rainfed areas, which was prepared through a process of several rounds of multi-disciplinary consultations and data analysis, provides a holistic approach to rainfed agricultural development in Tamil Nadu. Starting with a detailed analysis of the current status of rainfed farming in the state, it goes on to identify a set of feasible technological and institutional interventions that are realistic and economically viable in the existing socio-economic environment of rainfed areas. Location-specific technological interventions have been suggested for two hundred predominantly rainfed blocks along with costs of implementing these interventions. The third section deals with detailed methodology for selection of blocks and villages for implementing the plan, issues relating to implementation and monitoring of rainfed development project and the fourth section proposes an outline of expected outcomes and impact assessment.

Vision

Reduce poverty and enhance food security for the people in Tamil Nadu through sustainable development of rainfed agriculture

Mission

To improve the productivity and profitability of rainfed agricultural systems (crops, livestock, horticulture and agroforestry and pasture) using science-based interventions for augmenting water availability by vigorously promoting rainwater harvesting and its efficient and sustainable use by integrated watershed management approach while protecting the environment.

Objectives of the mission

- increasing the total productivity of rainfed farms by 40 per cent within five years
- increasing the total income of rainfed farmers significantly within five years and substantially minimizing their risk
- to introduce appropriate institutional interventions in rainfed areas so as to protect the rainfed farmers from risks, provide cheaper and timely credit and to enable better marketing of their produces; and
- to ensure sustainable environmental management in rainfed areas

Strategy

For enhancing incomes through sustainable development in rainfed areas, the proposed strategy is based on 4Es, i.e. Employment, Economic efficiency, Equity and Environmental protection.

To achieve the food security and improving the rural livelihoods, solution rests with the collective efforts of all the stakeholders, increased investments in research and development along with suitable policies and institutional changes. Moreover, to harness the existing potential of the rainfed agriculture to enhance the agricultural productivity significantly, substantial investments in rainfed areas are needed; knowledge driven management systems are also much needed along with the convergence of various programs operating in the areas. The three important pillars of this strategy are **collaboration** amongst the stakeholders, **convergence** of the activities and most importantly **empowerment** of the stakeholders through knowledge sharing and dissemination.

The strategy is to develop the five capitals i.e. natural resources, human resources, financial capital, social capital, and the physical facilities for achieving the goal by adopting the holistic approach through a consortium of the institutions with convergence of various activities. The strategy will be as depicted in figure to achieve **4Es** i.e., Equity, **E**fficiency, **E**mployment, and **E**nvironmental sustainability.

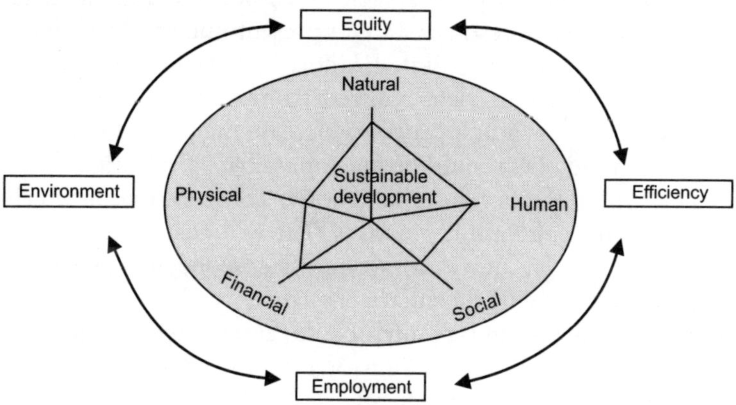

Development will be promoted through adoption of models integrating seamless integration of research and development. The knowledge and experiences of centres of excellence in the state, national and international levels will be brought together for achieving food security by managing scarce water resources efficiently and sustainably.

SWOT Analysis

A SWOT analysis is essential to comprehend the various dimensions of any system so as to plan the activities for scientific intervention to modify the system. This section provides a broad outline of the strengths, weaknesses, opportunities and threat in rainfed agricultural systems so as to provide a realistic scenario of the situation that will facilitate the preparation of realistic plans and programmes.

Strengths

Large areas of rainfed lands are available in almost all the districts of Tamil Nadu with diverse agro-climatic conditions that facilitate the cultivation of a variety of crops with different levels of moisture requirement. The soil types and other agro-ecological and socio-economic factors also show wider variation across different districts, thus, permitting a wide range of crop and related enterprises such as animal husbandry, agro-forestry and horticulture. These rainfed areas are endowed with different levels of productive factors such as labour and infrastructure facilities and these resources are at present underutilized due to frequent fallowing of rainfed lands. The yield gap between potential and actual yield realized from rainfed crops is very wide which offers ample scope to increase the yield closer to the potential yield. This could be made possible by the effective use of new technologies for soil and water conservation, new crop varieties and cultivation practices that will ensure better yield and profit under moisture stress. The scope for diversified farming system with subsidiary enterprises such as dairy, sheep and goat rearing and poultry could greatly reduce the risks associated with the cultivation of rainfed crops besides using the available resources efficiently.

Weaknesses

In spite of the existence of vast scope to improve productivity and profitability of rainfed agricultural production systems, the stagnation of yield is a major cause for concern. This is mainly because of very limited private investment in rainfed agriculture due to high risks associated with getting adequate returns to investment. The low average yield coupled with high variability makes rainfed crop production highly risky and un-remunerative for the farmers. Most rainfed crops are coarse cereals whose demand is declining and oilseed and pulse crops where yield and profitability are low. Given the low profitability of rainfed crops, higher share of cost of human labour and increasing scarcity of labour for agriculture is another hurdle in increasing the productivity and profitability of rainfed agriculture. Though mechanization could help reduce the cost of labour and ensure timeliness of agricultural operation under moisture stress, the small agricultural holdings offer very limited scope for mechanization and undertaking farm-level water conservation investments. Erosion proneness of the soils together with their poor physical and chemical properties in many rainfed areas is yet another major weakness in rainfed environments. Low investment capacity of rainfed farmers due to chronic poverty coupled with very limited flow of institutional credit is yet another weakness in rainfed agriculture. Further, in most of the rainfed areas there is a large scale diversification of livelihood away from agriculture with declining interest in agriculture. This is one of the major constraints in reviving the farmers' interest in rainfed agriculture and making them invest in productivity-enhancing technologies. Absence of effective insurance programme against risks associated with rainfed agriculture is a major institutional weakness in rainfed areas which acts as a major barrier in technology adoption.

Opportunities

The current yield levels of most of the rainfed crops are much lower than the potential yields, which offer ample scope for increasing the yield further. In view of the large area under rainfed agriculture and higher labour use for rainfed crops, there is scope for employment generation through increasing the area and productivity of rainfed crops per unit of land and labour. New institutional arrangements such as self-help groups in many of the rainfed areas can be successfully harnessed for channelising credit and insurance to rainfed areas. Provision of services such as credit, insurance and machinery rentals can be routed through self-help groups.

The increasing emphasis on watershed approach can be effectively utilized to implement soil and water conservation programmes in rainfed areas. The availability of modern soil and water conservation practices and communication and information technologies offer vast scope for speedy adoption of rainfed agricultural production technologies that will facilitate successful crop cultivation under moisture stress.

There is increasing consumer awareness about the importance of health foods such as fibre- and protein-rich crops such as coarse cereals and pulses which are the predominant crops in rainfed areas. Further, the growing industrial demand for some of the rainfed crops such as maize, cotton, sorghum, and sunflower is yet another opportunity to sustain the demand for rainfed agricultural outputs. The increasing awareness about the use of organically produced agricultural commodities could also be exploited in the rainfed areas to produce these crops at low cost.

Threats

Opportunities coexist with threats and this is more so in rainfed agricultural production. Shrinking resource base in some of the rainfed environments due to groundwater overexploitation and severe soil erosion caused by wind and rainfall is a potential threat in increasing the productivity of rainfed lands. The slow growth rate in prices of rainfed crops coupled with high risk in the yield of these crops is a major threat to sustain the profitability of those crops. Not only the failure of rainfall but also meager or back of rainfall during critical stages of crop growth leads to a complete failure of crops resulting in the collapse of farmers' livelihood in rainfed areas. In the absence of supplementary irrigation sources, there is limited scope for sustaining the household livelihood in drought years which compel the farmers to migrate to urban areas. Once a decision to migrate is taken it becomes irreversible and hence, migration is the most likely reason for increasing fallow lands in most of the rainfed areas. Frequent crop failure together with lack of appropriate insurance programmes make rainfed farming a highly risky venture for most of the poorer households.

Due to increasing opportunities for non-farm employment, the wages of agricultural labourers increase continuously while the profitability of rainfed crops fluctuate widely. Further, withdrawal of credit facilities by the credit institutions once the active government support is withdrawn is yet another threat to the survival of rainfed agriculture. Any rainfed agricultural development project should, therefore, consider all these strengths, weakness, opportunities and threats before embarking on devising and implementing rainfed agricultural development programmes.

I
SITUATIONAL ANALYSIS
A DETAILED ACCOUNT OF THE CURRENT STATUS OF RAINFED AGRICULTURE IN TAMIL NADU

Land Use Pattern

Rainfed areas are characterized by poor water resources due to low rainfall with associated problems such as low vegetative cover in the form of cultivated crops or forests, high levels of soil erosion, and vast areas of fallow lands. Cropping intensity is also comparatively lower in dry lands due to limited scope for double cropping. Details on land use pattern and cropping intensity in rainfed districts are presented in Table A1 and A2. Fallow lands, especially other fallow lands which remain uncultivated for more than one year are higher in most of the rainfed districts than the state average. The share of other fallow lands to total geographical area was higher in the central and southern districts of the state accounting for 15 to 25 per cent of the geographical area of these districts. Fallowing of lands is mainly due to the inadequate water availability in rainfed areas.

The location coefficients for fallow lands which indicate the concentration of fallow lands are presented in Table 1. Increasing trend in location coefficient (which indicates the concentration of fallow lands) in Tiruchirappalli, undivided Ramanathapuram and Pudukkottai districts is a major impediment in stabilizing the gross cropped area in these districts. Provision of supplementary sources of irrigation is essential to halt the increasing trend in fallow lands in many parts of the state.

Table 1. Concentration of Fallow Lands in Selected Districts of Tamil Nadu

S.No.	Districts	Location coefficient for fallow lands			
		1970-71	1980-81	1990-91	1999-2000
1.	North Arcot	0.91	0.87	0.80	0.48
2.	Coimbatore	1.20	0.56	0.52	0.56
3.	Salem	1.01	0.90	0.49	0.35
4.	Dharmapuri	0.23	0.52	0.14	0.14
5.	Tiruchirapalli	0.85	0.68	2.21	1.53
6.	Pudukottai	0.97	1.37	1.08	2.05
7.	Madurai	0.78	0.63	0.85	0.76
8.	Ramanathapuram	1.10	1.99	2.10	2.23

Land Holding Structure

The details on land holding structure in predominantly rainfed districts is provided in Tables A-3 – A-5. The data show that the marginal farms account for more than 60 per cent of the total number of holdings in all these districts except in Coimbatore and Erode districts. However, their share in total operated area was less than 40 per cent in most of these districts indicating unequal distribution of land. In 11 out of 17 districts the average size of holdings was less than one hectare.

Rainfall

A good spatial and temporal distribution of rainfall is a pre-requisite for the growth of agricultural sector. However, the state of Tamil Nadu faces frequent droughts and years of deficit rainfall. The rainfall pattern in the state shows wide fluctuation across space and time. The wide temporal variability is the major factor responsible for high yield risk in most of the rainfed crops. Though the long-term average over the last 40 years is around 900 mm which is close to the normal rainfall of the state, year-to-year variation in total amount of rainfall and skewed distribution of rainy days in a year are the major impediments in improving the productivity of rainfed crops. In the absence of a well-organized crop insurance scheme, farmers in rainfed areas face extremely high fluctuations in their income, thus, making them vulnerable to frequent loss in farm business. The long-term coefficient of variation in the state shows that there is about 10 to 20 per cent deviation in normal rainfall

(Table 2). The district-wise analysis of rainfall (Table 3) reveals that most of the predominantly rainfed districts show higher variability in rainfall than the state-level variability.

Table 2. Rainfall Variability in Tamil Nadu

	Mean	Standard deviation	Coefficient of variation
1960s	928.22	136.03	14.66
1970s	931.53	133.92	14.38
1980s	880.34	174.06	19.77
1990s	917.23	103.14	11.24
2000s	836.55	135.02	16.14
1960-2000	907.26	135.51	14.94

Rainfall trend in Tamil Nadu

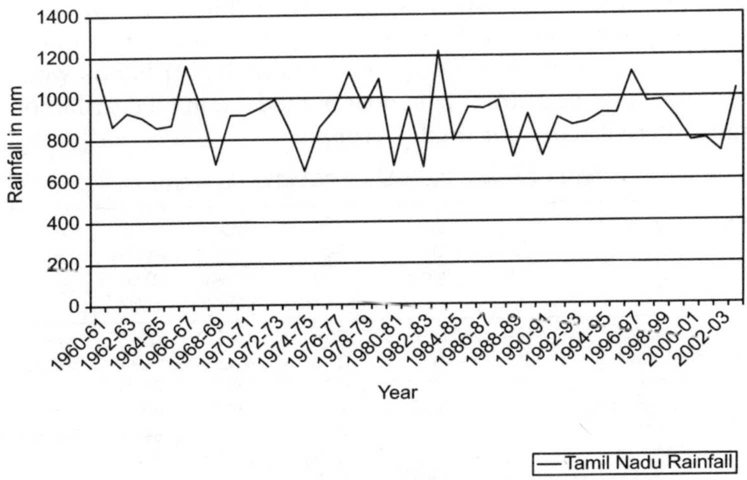

Table 3. Rainfall Variability in Predominantly Rainfed Districts (1960-2000)

S.No.	Districts	Mean (mm)	Standard deviation	Coefficient of variation (%)
1	Vellore	961.80	215.99	22.46
2	Cuddalore	1121.25	266.85	23.82
3	Villuppuram	1107.42	257	24.00
4	Tiruvannamalai	965.91	314.80	32.59
5	Dharmapuri	862.42	198.59	23.03
6	Salem	857.49	149.40	17.42
7	Namakkal	626.42	140.57	22.44
8	Coimbatore	667.64	163.72	24.52
9	Erode	696.55	122.80	17.63
10	Trichy	778.50	167.19	21.48

11	Karur	508.25	152.54	30.01
12	Perambalur	803.33	296.66	36.93
13	Madurai	829.91	165.75	19.97
14	Theni	867.13	186.47	21.50
15	Dindigul	838.18	197.60	23.57
16	Ramanathapuram	805.07	163.03	20.25
17	Pudukkottai	813.72	194.30	23.88
18	Sivagangai	868.62	121.40	13.98
19	Thanjavur	1159.00	284.64	24.54
20	Thoothukkudi	665.85	177.10	26.60
21	Tirunelveli	852.67	372.05	43.63
22	Virudhunagar	770.36	239.61	31.10
	Tamil Nadu	907.26	135.51	14.94

Rainfall and Length of Growing Period

Season-wise normal rainfall and length of growing period in each of the 22 districts selected for the implementation of the mission are provided in Table 4. The figures indicate that out of 22 districts only seven have uni-modal rainfall distribution while the remaining 15 districts fall under bi-modal rainfall distribution thus offering scope for double cropping.

Table 4. Seasonal Distribution of Rainfall and its Pattern

S.No.	Districts	Jan-Feb	Mar-May	Jun-Sep	Oct-Dec	Total annual rainfall (mm)	Type of rainfall
		Winter	Summer	Southwest monsoon	Northeast monsoon		
1	Cuddalore	56.4	89.3	373.6	716.5	1235.8	Bi-modal
2	Villupuram	34.5	77.1	433.0	484.8	1029.4	Bi-modal
3	Vellore	20.3	101.7	442.0	353.0	917.0	Bi-modal
4	Thiruvannamalai	32.8	108.2	465.8	439.8	1046.6	Bi-modal
5	Salem	21.3	149.7	380.0	347.0	898.0	Bi-modal
6	Namakkal	18.1	150.4	317.0	291.0	776.5	Bi-modal
7	Dharmapuri	14.0	146.7	391.8	303.4	855.9	Bi-modal
8	Tiruchirapalli	25.0	110.1	270.3	356.1	761.5	Bi-modal
9	Karur	24.0	103.1	249.7	365.4	742.2	Bi-modal
10	Perambalur	34.5	115.9	349.6	449.6	949.6	Bi-modal
11	Pudukottai	38.2	114.6	350.7	418.0	921.5	Bi-modal
12	Thanjavur	50.7	114.6	342.0	545.7	1053.0	Bi-modal
13	Madurai	29.8	131.8	305.4	373.0	840.0	Bi-modal
14	Dindigul	33.0	148.0	251.4	399.2	831.6	Bi-modal
15	Sivagangai	35.8	135.1	289.6	415.5	876.0	Bi-modal
16	Tirunelveli	72.6	141.9	92.6	429.8	736.9	Uni-modal
17	Thoothukudi	46.6	112.2	86.8	410.1	655.7	Uni-modal
18	Theni	48.4	222.7	178.4	384.0	833.5	Uni-modal
19	Coimbatore	26.1	148.4	192.9	327.0	694.4	Uni-modal
20	Erode	20.7	154.1	213.1	323.5	711.4	Uni-modal
21	Ramanathapuram	53.9	123.8	136.1	507.4	821.2	Uni-modal
22	Virudhunagar	42.0	174.6	181.8	431.2	829.6	Uni-modal

Soil Resources of Rainfed Districts

The type of soils found in predominantly rainfed districts are given in the following table. Red loam and black soil are the major types of soils in most of the rainfed districts.

Table 5. Soil Types in Rainfed Districts

Type of soil	Areas
Red loam	Parts of Salem, Dharmapuri, Coimbatore, Tiruchirappalli, Madurai, Ramanathapuram, Sivagangai, Thoothukudi, Virudhunagar, Dindigul districts.
Black soil	Parts of Vellore, Thiruvannamalai, Salem, Dharmapuri, Madurai, Ramanathapuram, Sivagangai, Thoothukudi, Virudhunagar and Dindigul districts.
Sandy coastal alluviam	On the coasts of Ramanathapuram district
Red sandy soil	Small patches in the districts of Coimbatore

Irrigation sources

The data on irrigation infrastructure and the share of rainfed area to total cropped area are presented in Table 6. In most of the districts, the share of gross rainfed cropped area to the gross cropped area was higher than 50 per cent. In districts such as Dharmapuri, Perambalur, Thoothukkudi, Dindigul, Ramanathapuram, Salem and Namakkal, the share of gross rainfed area was higher than 60 per cent of the gross cropped area. A simple correlation analysis of these data reveal that number of wells per ha of net sown area had significant negative correlation with the extent of rainfed area (-0.22), and significant positive correlation with irrigation intensity (0.76) and cropping intensity (0.60) indicating that private groundwater development plays a crucial role in mitigating the extent of rainfed cultivation and increasing the irrigation intensity and cropping intensity. Similarly, there was a very strong positive cross-sectional (across districts) correlation between irrigation intensity and cropping intensity (0.90) indicating that irrigation has strong impact in increasing the gross cropped area. Both the number of wells per ha and the number of electric pumpsets per ha of net sown area were very less in the chronically drought-prone districts such as Ramanathapuram, Sivagangai, Thoothukkudi, Perambalur, Dharmapuri and Pudukottai. Therefore, there is significant potential to develop groundwater irrigation in many parts of these districts so as to increase the area under cultivation and to stabilize the existing area under irrigation. The districts with predominantly tank irrigation such as Ramanathapuram, Sivagangai, and Pudukkottai have substantial potential for groundwater development.

Table 6. Irrigation Infrastructure and Cropping Intensity

S.No.	Districts	No.of electric pumpsets per ha of net sown area	Number of wells per ha of net sown area	% of Gross rainfed area to gross cropped area (2001-02 to 2003-04)	Irrigation intensity (2001-02 to 2003-04) (%)	Cropping intensity (2001-02 to 2003-04) (%)
1	Cuddalore	0.28	0.20	61.43	117.05	128.97
2	Villuppuram	0.47	0.61	55.23	120.76	115.16
3	Vellore	0.63	0.66	51.79	128.94	120.64
4	Tiruvannamalai	0.67	0.87	41.14	137.59	126.50
5	Salem	0.58	0.59	63.61	131.13	131.06
6	Namakkal	0.32	0.53	66.37	142.79	137.72
7	Dharmapuri	0.25	0.39	72.21	117.75	115.13
8	Coimbatore	0.36	0.39	46.72	104.99	104.21
9	Erode	0.46	0.44	52.48	111.65	105.62
10	Tiruchirappalli	0.80	0.53	43.51	114.27	108.07
11	Karur	0.43	0.57	53.94	102.96	101.36
12	Thanjavur	0.16	0.20	79.97	121.66	126.54
13	Perambalur	0.01	0.29	71.34	109.47	107.30
14	Pudukkottai	0.27	0.27	36.59	101.82	101.46
15	Madurai	0.37	0.36	44.13	107.83	104.72
16	Theni	0.29	0.26	52.48	114.15	107.98
17	Dindigul	0.35	0.39	65.22	107.93	102.62
18	Ramanathapuram	0.04	0.04	62.63	100.00	100.00
19	Virudhunagar	0.24	0.26	57.02	107.29	103.28
20	Sivagangai	0.17	0.18	39.34	100.12	114.92
21	Tirunelveli	0.49	0.64	69.76	121.47	116.50
22	Thoothukkudi	0.17	0.21	76.39	105.37	102.14
	STATE	0.34	0.38	49.27	117.29	116.13

Human resources, literacy and poverty profile

The occupational classification of workforce, poverty profile and decadal population growth in the predominantly rainfed districts are provided in Table A-6 to A-8. The data indicate that the share of cultivators to total workforce in Tamil Nadu has been less compared to that of India. Similar trend is found in the share of agricultural labour to total workforce as well. Rural poverty does not appear to be correlated with the extent of rainfed farming possibly due to migration and the availability of non-farm employment opportunities in rainfed areas.

The decadal change in population reveals that in the central and southern districts, the growth in total population has been less than the state level growth of about 11 per cent. Tamil Nadu state is in the forefront of urbanization among all the states in India with about 45 per cent of total population living in urban areas. The urban population has increased by about 44 per cent during the last decade while rural population has declined by about 5 per cent during this period. Reflecting this state-wide trend, rural population has been declining in 16 out of 22 districts selected for the mission. In nine out of the 16 districts, the decline in rural population has been higher than the state-level decline in rural population. Moreover, the share of farmers to total work force has been declining at 2.67 per cent per annum during the 1990s. Given the very low income in rainfed agriculture, many farm households in rainfed areas have diversified their income portfolio away from agriculture, which is reflected in higher rate of migration from rural areas as shown by the declining trend in rural population. This is evident from very high rates of urban population growth and negative growth or very low positive growth of rural population in the entire state of Tamil Nadu and in all the rainfed districts.

Livestock

The data on livestock population per unit land area are presented in Table A-9. The data show that there are sizable number of livestock per ha of net sown area though it shows a wide variation across districts ranging from 2.61 per ha in Ramanathapuram to 9.39 per ha in Dindigul district. The livestock population density per ha of geographical area varies from 1.17 per ha in Ramanathapuram to 3.42 per ha in Dindigul. A break-up of total livestock into large cattle (cows and buffaloes) and sheep and goats shows that the density of both of these two groups ranged from 0.5 to 2.0 across different districts. In general, the livestock population density appears to be less in low rainfall areas such as Coimbatore district and in all four southern districts when compared to the other districts and the state average. Scarcity of fodder during drought years might be an important reason for less number of livestock in the dry regions. Hence, there is scope for promotion of livestock-based livelihood system in these districts through appropriate mechanisms for overcoming fodder scarcity in drought years. Creation of community fodder reserves/fodder buffer stocks is one possible arrangement to circumvent this problem. Protection of community grazing lands and development of common property wastelands through pasture development would go a long way in promoting livestock development in these areas.

Farm machinery

The availability important farm implements and machinery per 1000 ha of net sown area are presented in Table A-10. The data reveal that the number of seed drills, seed-cum-fertilizer drills, tractors and power tillers which are important labour-saving equipments required in rainfed conditions are less than the state average in most of the rainfed districts.

Infrastructure

(a) Banking and agricultural insurance

The development of banking infrastructure is fairly adequate in most of the districts while that of agricultural insurance has been uniformly poor (Table A-11 and A-12). In fact, the banking network in terms of population served per bank is better in the chronically drought prone southern districts as compared to northern districts. The per cent of rural branches to total number of branches in the respective districts also appears to be higher than the state average in most of the rainfed districts. But the extent to which rainfed farmers are benefited by the nationalized banks remains questionable since banking institutions are reluctant to provide credit for cultivating rainfed crops. Agricultural insurance remains a distant dream in most of the districts and this is more so in rainfed districts. Groundnut is the only rainfed crop covered under crop insurance, but the actual coverage in terms of number of farmers is abysmally low under unirrigated areas as compared to irrigated areas.

(b) Roads and markets

The data on roads and markets presented (Table A-13) reveal that the rainfed districts have fairly better network of regulated markets and rural godowns as compared to the state average. Rainfed districts account for 178 of the 273 regulated markets in the state 70 rural godowns compared to 108 in the state as a whole. But arrivals and sales through regulated markets are very meager and in many of the regulated markets in predominantly rainfed districts such as Ramanathapuram, Virudhunagar, and Dharmapuri, it is practically nil. The average road length per square kilometer of geographical area is very much close to the state level average of 0.43 kilometer in most of the districts.

Role and Importance of Rainfed Agriculture

The trends in area under cultivation in the state are presented in Figures 2 and 3. The area under rainfed cultivation appears to be declining in Tamil Nadu over the period of 44 years along with the declining trend in net sown area and gross cropped area, even though the rate of decline is very slow. The net area under rainfed cultivation has declined from 3.5 lakh ha to 2.5 lakh ha while the gross area under rainfed cultivation has declined from about 4.00 lakh ha to 2.90 lakh ha. This only reflects the general trend in the decline in area under cultivation and the increase in area under fallow lands and land put to non-agricultural uses. The share of net sown area under rainfed cultivation to total net sown area has only marginally declined from about 59 per cent to 54 per cent in spite of the sustained efforts to bring more area under cultivation through several public irrigation projects and private initiatives in groundwater development. This is reinforced by the fact that both the net irrigated area and gross irrigated area in the state have stagnated over the last 44 years. The accelerated public and private investments to develop large scale surface irrigation projects and groundwater development have been more than offset by the continuous declining trend in community based tank system which are concentrated more in drought prone areas such as southern districts. Therefore, sustained efforts are necessary to conserve and protect these local irrigation systems besides developing new, large-scale projects.

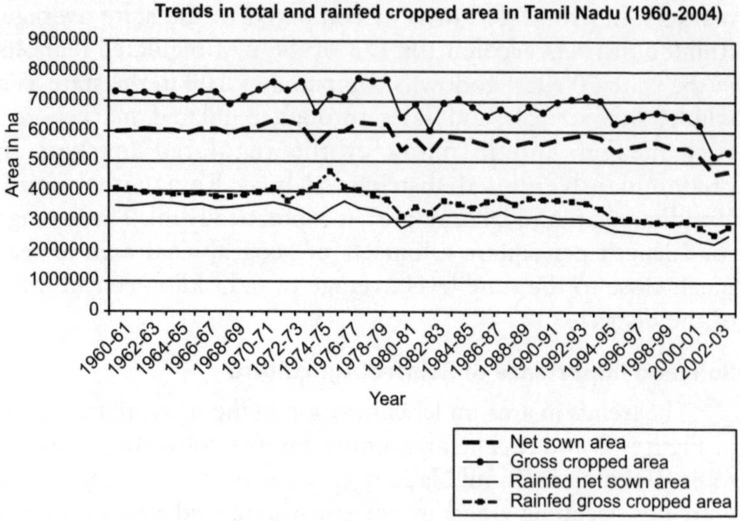

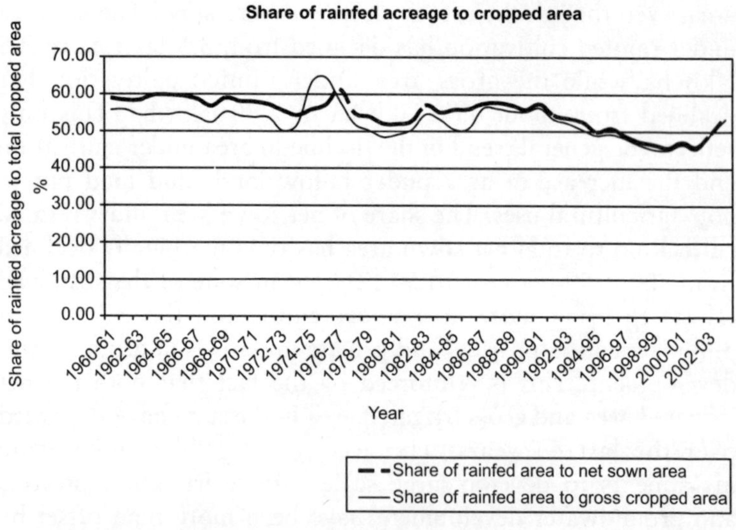

Quantity and value of rainfed crop output and employment generation

Rainfed crop production plays a significant role in the livelihood of rural people. This is evident from the share of rainfed production, value of rainfed crop production and total employment generated by rainfed cultivation. In case of sorghum, bajra, ragi, cotton, groundnut and gingelly, both the rainfed crop output and the value of production on the average account for about 70 per cent of the total output and the value of output (Tables 7-9).

Table 7. Share of Rainfed Agriculture in Total Production (Average for triennium ending 2002-2003) (Lakh tonnes)

Crops	Irrigated	Rainfed	Total
Rice	30.47 (94.55)	1.76 (5.45)	32.23(100)
Sorghum	0.23 (9.22)	2.23 (90.78)	2.46 (100)
Bajra	0.10 (5.65)	1.63 (94.35)	1.73 (100)
Ragi	0.49 (27.94)	1.27 (72.06)	1.76 (100)
Cotton (lint) 170 kg bale	0.82 (67.13)	0.40 (32.87)	1.22 (100)
Groundnut	4.12 (44.68)	5.06 (55.14)	9.18 (100)
Gingelly	0.09 (29.76)	0.20 (70.24)	0.29 (100)

Note: Figures in parentheses denote percentage to total

Table 8. Value of Production in Rainfed Agriculture (Average for triennium ending 2002-2003) (Rs.crores)

State	Irrigated	Rainfed	Total
Rice	1615.98 (94.55)	93.18 (5.45)	1709.17 (100)
Cholam	12.74 (9.20)	125.47 (90.80)	138.21 (100)
Cumbu	4.75 (5.70)	79.29 (94.30)	84.04 (100)
Ragi	28.00 (27.90)	72.23 (72.21)	100.23 (100)
Cotton	158.35 (67.10)	77.53 (32.90)	235.88 (100)
Groundnut	633.51 (44.90)	778.77 (55.10)	1412.27 (100)
Gingelly	18.30 (29.80)	43.20 (70.20)	61.50 (100)

Note: Figures in parentheses denote percentage to total

Table 9. Employment Generation in Rainfed Agriculture (Average for triennium ending 2002-2003)

Crops	Rainfed Area (lakh ha)	Employment	
		Mandays / Ha	Total employment (lakh mandays)
Paddy	1.35 (4.30)	180.57	243.77 (8.46)
Black gram	1.65 (5.25)	63.2	104.28 (3.62)
Cotton	0.54 (1.72)	92.83	150.13 (1.74)
Cumbu	1.52 (4.84)	69.35	105.41 (3.66)
Green gram	1.10 (3.50)	46.2	50.82 (1.76)
Groundnut	4.20 (13.38)	147.54	619.67 (21.51)
Maize	1.12 (3.57)	109.02	122.10 (4.24)
Ragi	0.98 (3.12)	58.95	57.77 (2.01)
Redgram	0.44 (1.40)	105.33	46.35 (1.61)
Gingelly	0.64 (2.04)	64.58	41.33 (1.43)
Sorghum	3.82 (12.17)	71.53	273.24 (9.49)
State	31.40 (100)	91.74	2880.52 (100)

Note: Figures in parentheses denote percentage to state total

Problems in rainfed agriculture

Yield performance of rainfed crops

The performance of rainfed agriculture has been poor in terms of improvement in both technology adoption and productivity. The yields of most of the rainfed crops have remained nearly stagnant over the last 40 years as indicated by the data presented in Table 10.

Table 10. Trends in yield of predominantly rainfed crops

	1960s	1970s	1980s	1990s	2000
Sorghum	748	775	611	927	824
Bajra	608	667	614	973	1169
Samai	454	481	504	881	809
Varagu	1048	1090	1026	1183	1302
Redgram	460	462	551	554	677
Blackgram	331	319	352	454	420
Cotton	142	202	64	214	231
Groundnut	1101	906	778	1183	1477
Gingelly	317	321	283	318	385

Source: Season and Crops Report for Tamil Nadu (Various issues)

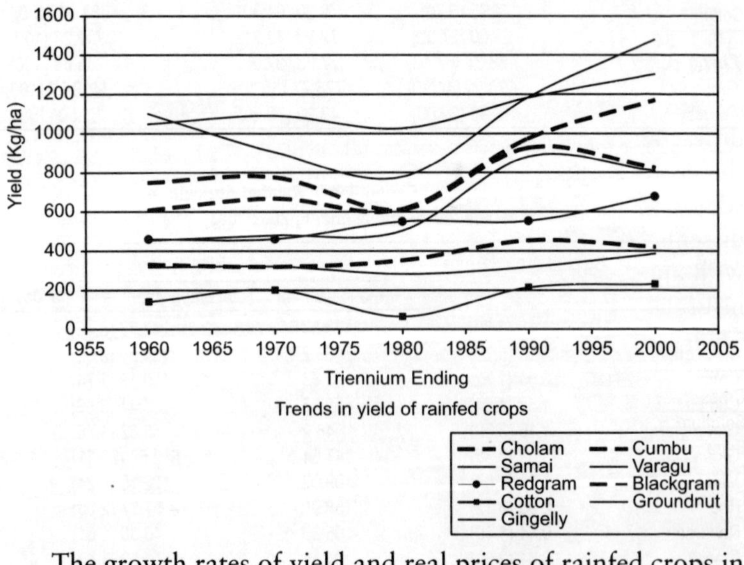

Trends in yield of rainfed crops

The growth rates of yield and real prices of rainfed crops in the state are presented in Table 11. Most of these rainfed crops have recorded a very low or negative growth rate in yield, while the real prices of most of these crops have declined over the last two decades. The decline in rainfed crop prices is a major disincentive for the adoption of modern technologies in these areas that makes rainfed farming economically unsustainable in the long run. This possibly is a reason behind the large-scale out-migration from rainfed areas.

Table 11. Growth Rate of Yield and Real Prices of Major Rainfed Crops in the State

Crops	Yield			Real prices		
	1980s	1990s	Overall	1980s	1990s	Overall
Sorghum	5.85	-0.73	1.74	-2.90	1.05	-1.28
Bajra	4.20	3.09	2.48	-3.14	0.72	-1.20
Ragi	3.99	0.20	2.29	-2.86	0.79	-1.09
Maize	1.00	-0.11	0.23	-0.71	-0.50	-0.83
Red gram	2.02	1.13	0.18	-2.24	1.00	0.33
Green gram	4.38	-0.06	2.08	-2.33	3.11	0.42
Bengal gram	1.75	0.24	0.40	0.17	2.92	1.23
Chillies	0.70	-0.05	-0.89	1.70	2.28	-0.73
Coriander	-0.32	0.67	-3.28	-1.57	-2.29	-1.23
Groundnut	2.68	3.47	3.32	-3.15	-3.39	-1.89
Gingelly	-2.20	4.77	2.22	-3.07	-3.63	-2.18
Castor	-4.14	0.01	-1.65	-2.7	1.57	0.61
Cotton	4.50	0.20	0.93	-1.60	2.60	-0.80

Yield difference between irrigated and rainfed crops

Consequent to low levels of technology adoption, the yield difference between rainfed and irrigated condition for most of the crops is more than 75 per cent (Table 12). There is a vast scope to bridge at least a part of this gap through the dissemination of low-cost, risk mitigating technologies along with appropriate institutional support in the form of credit and insurance.

Table 12. Yield difference between Irrigated and Unirrigated Situations

Crops	1980	1990	1980-2003
Sorghum	979	990	986 (121.17)
Bajra	1356	1434	1402 (149.39)
Ragi	1260	99	524 (25.81)
Cotton	347	210	267 (166.08)
Groundnut	714	1002	882 (77.54)
Gingelly	268	442	369 (102.05)

Note: Figures in parentheses indicate percentage yield difference between irrigated and unirrigated situations

Low level of technology adoption

Adoption of modern technologies is generally very limited in rainfed production environments mainly because of high risks associated with wide fluctuations in rainfall and lack of irrigation which is the leading input in agriculture.

Technological Gap in Awareness and Adoption of Dryland Practices

A few studies on the extent of adoption of modern technologies in rainfed areas conducted by the Tamil Nadu Agricultural University were reviewed to find out the extent of adoption and the reasons for low rate adoption. As majority of the farming community own lands with scarce water resource and with sole dependence on rainfall, there are a number of specially developed rainfed technologies. However, intensive as well as extensive methods of technology transfer have been rather slow in these areas resulting in a wide gap in technology adoption and also in awareness. Studies on extent of technology adoption in rainfed areas reveal mixed trends. Adoption rate of rainfed cultivation technologies in Tamil Nadu for various crops were very much discouraging though certain rainfed practices recorded 70-80% adoption.

1. Considering the extent of adoption of maize and sorghum in dry tracts of Dindigul District, 82.22 per cent of maize growers and 66.67 per cent of sorghum growers have fully adopted summer ploughing to eradicate weeds. In seed rate, a maximum number of maize growers (88.89%) had gone for full adoption, whereas only 17.78 per cent of sorghum cultivators adopted the recommended seed rate. The gap is so wide due to the reason that the germination is low in rainfed tract. Due to non-availability of bio-fertilizer, only a meagre proportion (8-10%) have gone for application of biofertilizers for both maize and sorghum.

Similarly, due to lack of awareness and knowledge, spacing to be adopted and micro-nutrient to be applied reflected a very poor adoption rate in both the crops. A wide gap is found in the above practices. It is disheartening to note that there was 100 per cent non-adoption in disease management.

2. Studies in eastern and western ghat areas on technology-wise adoption of indigenous in-situ on farm conservation technologies for rainfed ragi crop show that only 40-50 per cent adoption was observed in indigenous technologies namely sowing of seeds, raised bed nursery, application of cow dung, raising crops at times of good rain, crop rotation, and few other technologies. Here too, a larger proportion of farmers have not gone for adoption of improved technologies in ragi.

3. Adoption of specific dryland technologies such as drought tolerant varieties, advance sowing, seed treatment, intercropping and mixed cropping were adopted by less than 70 per cent of the farmers. Application of recommended dose of chemical fertilizers, weeding and fallowing were adopted only by a meagre proportion of farmers ranging from 20-25 per cent. So also, a lesser percentage (less than 25 per cent) of adoption was observed in summer ploughing, mulching, farm ponds, contour bunding, broad bed and furrow, mid-term correction and the like. A vast gap is observed in the adoption of these important dryland technologies.

4. While looking into the awareness of dryland technologies in rainfed cotton and pulses, a study conducted in Salem district and certain pockets of Aruppukkottai taluk in Virudhunagar district showed that only 50 per cent of the respondent farmers were aware of the seed treatment. Only 50 per cent of the farmers were found to be aware of the application of azospirllum. About 75 per cent of the respondents were aware of broad-bed and furrow system, the modern cotton varieties such as LRA 5166 and MCU –10. Other practices like weedicide application, use of pheromone traps, Rhizobium culture for seed treatment in pulses, and foliar application of fertilizer were known to a very low percentage (less than 20 per cent) of farmers. A wide gap was observed even at the awareness stage of technology dissemination.

A study on knowledge and extent of adoption of integrated weed management in Erode and Coimbatore districts, reveals that the major constraints in adoption of modern technology were labour scarcity, poor quality of inputs etc. Another study on awareness and adoption of recommended eco-friendly rainfed agricultural practices for cotton in Salem District revealed that labour scarcity, lack of assured irrigation, lack of technical guidance, lack of knowledge to identify predators, difficulty in maintaining pheromone traps were the major constraints in the adoption of these technologies.

Is low input use a cause for low yield of rainfed crops?

Despite the growing use of fertilizer in rainfed areas, soil fertility is a major problem. Use of organic manures is inadequate and many plots do not receive fertilizer or manure over a period of several years of continual cropping. Chemical fertilizer is still only used on a little over half of the rainfed grossed cropped area. Use of fertilizer is related to rainfall, access to credit and proximity to a town. Use of HYVs and an efficient fertilizer distribution system may partly compensate for lower response to fertilizer in drier areas as shown by the example of Gujarat where most farmers use modest doses of fertilizer even in dry conditions. Fertilizer use is also generally much more widespread on cash crops compared to food grains. Indeed it is frequently observed that fertilizer use varies considerably within a farm, with many farmers applying fertilizer to a cash crop such as cotton but not to a food crop. This, of course, reflects the relative profitability of fertilizer use and the fact that fertilizer use on food grains is often not very profitable

In view of the poor soil nutrient status in most of the rainfed areas, fertilizer is a key input in enhancing the productivity of rainfed crops. However, water is the complementary input which limits the scope for application of fertilizers. Therefore, the fertilizer use for rainfed crops is comparatively much lower than the irrigated crops. Consequently, yield levels of most of the rainfed crops are much less than the potential yield. The yield of most of the dryland crops shows a very slow long-term growth (Table 13). Therefore, low input use appears to be an important factor responsible for low yield of rainfed crops.

High risk in rainfed agriculture

Rainfed agriculture, by nature, is fraught with the problem of high risk due to wide fluctuation in rainfall. In the absence of supplemental irrigation, yield risk in rainfed cultivation is directly proportional to rainfall variability. The data provided in Table 14 reveal that the yield risk is substantial in most of the rainfed crops. Hence, risk mitigation through both technological and institutional interventions is a prerequisite to make rainfed farming viable and sustainable.

Table 13. Fertilizer Use and Yield of Rainfed Crops

(Yield in quintals per ha; Fertilizer use in kg/ha)

Crop		1999-00
Jowar	Yield	8.15
	Fertilizer	32.65
Bajra	Yield	12.36
	Fertilizer	46.32
Black gram	Yield	7.9
	Fertilizer	24.68
Redgram	Yield	4.16
	Fertilizer	8.36
Groundnut	Yield	16.32
	Fertilizer	149.35
Gingelly	Yield	4.12
	Fertilizer	45.26
Cotton	Yield	11.26
	Fertilizer	69.14

Source: Farm survey data, Comprehensive scheme on cost of cultivation of principal crops, Department of Agricultural Economics, TNAU, Coimbatore

Table 14. Yield Variability in Rainfed Crops (Coefficient of variation)

Crops	Coefficient of variation in yield (%)		
	1980	1990	1980-2003
Sorghum	24.17	15.93	19.38
Bajra	20.12	12.33	19.46
Ragi	27.80	156.93	155.74
Cotton	47.27	24.49	39.08
Groundnut	31.10	18.89	25.95
Gingelly	21.67	50.11	43.63
Bengal gram	6.50	4.76	5.48
Red gram	23.39	15.55	19.09
Green gram	19.33	11.01	16.89
Black gram	16.24	12.66	15.28
Horse gram	23.27	20.50	22.18
Castor	13.30	2.81	12.43
Coriander	–	45.26	43.19
Sunflower	–	25.32	37.26

Low profitability of rainfed crops

Profitability of rainfed crops has not shown significant improvement over the last 20 years, which is another major cause for concern in sustaining the farmers' interest in rainfed cultivation (Table 15). Given the average land holding size of about one hectare, the income from rainfed cultivation which is in the order of Rs.2500 to Rs.10000 per ha is grossly inadequate to sustain the household economy. Further, higher risks in rainfed cultivation together with low profitability makes it unattractive proposition unless the farm households in rainfed areas have access to supplementary sources of income either on farm or off the farm. Increasing the profitability of rainfed crops with reduced risk is the only option available to make rainfed farming viable. A long-term possibility is to concentrate on breeding crop varieties for drought tolerance, while in the short run, efforts on soil and water conservation is likely to play a major role in enhancing the productivity of rainfed crops. Reduction of cost of cultivation of rainfed crops is another low cost strategy to increase the profitability in the short-term. Given the higher share human labour accounting for more than 50 per cent of total cost of cultivation of most of the rainfed crops (Table 16), use of labour-saving implements and machinery is an important strategy to reduce the cost of cultivation of rainfed crops.

Table 15. Low Profitability in Rainfed Cultivation

Crop	Net income from rainfed crops (Rs./ha) at current prices		
	1984-86	1990-92	2003-05
Sorghum	1183	3741	2474
Bajra	–455	3141	3579
Maize	NA	5888	2822
Red gram	893	3880	NA
Black gram	1842	6296	4799
Green gram	1931	2815	5825
Cotton	882	NA	NA
Groundnut	–586	5824	10815
Gingelly	1212	5544	9219
Sunflower	2759	2467	3893
Chillies	NA	NA	–381

Source: Farm survey data, Comprehensive scheme on cost of cultivation of principal crops, Department of Agricultural Economics, TNAU, Coimbatore.

Based on the above analysis, the problems facing rainfed agriculture could be summarized as below:

- Low mean rainfall with high variability around mean indicating higher weather risk
- Resource constraints—water scarcity, poor nutrient status of soils and soil erosion

Table 16. Share of Human Labour in total Cost of Cultivation of Rainfed Crops (% to total cost)

Crops	1999-00
Sorghum	57.45
Bajra	62.33
Ragi	62.03
Black gram	42.09
Redgram	67.30
Cotton	54.27
Groundnut	42.43
Gingelly	8.84

Source: Farm survey data, Comprehensive scheme on cost of cultivation of principal crops, Department of Agricultural Economics, TNAU, Coimbatore

- Low application of modern inputs such as fertilizers and pesticides and low productivity
- High risk in crop production
- Small and uneconomic size of holdings
- Lack of institutional support–credit, insurance, and infrastructure
- Chronic poverty and low investment on both crop production and resource conservation activities.
- Stagnation in dryland technology
- Livelihood diversification away from agriculture resulting in abandoning of agriculture or increase in fallow lands
- Low off-take of soil and water conservation technologies in rainfed areas due to
  - Relatively high cost of the package,
  - Draft power constraints, and
  - Lack of institutional support

Thus, there is a kind of vicious cycle in rainfed agriculture with low rainfall, poor resource base and inadequate credit and insurance markets on the one side and low investment, low levels of technology adoption and low yield and income on the other side.

Impact Assessment of Dryland Development Programme

Any investment must generate adequate pay-off both monetary and non-monetary. Unless intended output/benefits are realized, there is no incentive for further investment and development intervention. Impact study will explain what has been done in a project, how things have been done, what has been achieved, and what future prospects or options may exist in terms of sustainability and, possibly, replicability. This means that one must seek to link **past** status and events, to **present** status and events, by specifying any differences between them and exploring how such differences have been created. Additionally, if one is looking into future, one must ask whether observed changes are likely to continue and what factors are likely to influence continuation, positively and negatively. Also, one of the purposes of evaluations is to promote learning. Learning can occur only when we know why things have happened the way they have.

The overall objective of improving dry land agriculture is poverty alleviation through sustainable agriculture practices. The specific objectives are:

1. Effective use of available land resources;
2. Increasing employment;
3. Increasing income from crop and livestock activities; and
4. Putting in place an organization to facilitate viable and sustainable dryland agriculture

Typically, the impact assessment of rainfed agricultural development project will involve the following steps:

- Identification of monitoring indicators under the following impact categories:
  - Technological improvement

    - Economic improvement
    - Social and institutional improvement
    - Environmental improvement
- Collection of information to quantify the indicators identified above
- Estimation of impact indicators
- Summary of achievements of projects vis-à-vis initial goals
- Lessons learnt for future up-scaling/strategies for horizontal diffusion

Policies for Rainfed Agricultural Development

Rainfed and irrigated agriculture coexist in practically every village in Tamil Nadu. Public investment programs, however, usually cannot be targeted so precisely. For practical purposes, they need to be planned and implemented on a larger scale, such as at the village, taluk, district or state level. For example, public programs that provide credit, employ people, or build roads cannot target their efforts to either rainfed or irrigated agriculture; they can only target areas that are relatively more irrigated or more rainfed. Price policies, on the other hand, can attempt to target rainfed or irrigated agriculture within a given location by targeting crops that may be more likely to be rainfed or irrigated. But few crops are either 100% irrigated or 100% dryland, so some spillover will always remain. Also, every crop is grown over a large geographic area, so it is difficult to isolate the socioeconomic and agroclimatic variables affecting their performance. And while price policies are important, they are not the only approach through which policy makers can influence agricultural development.

(I) Dryland areas with good groundwater potential

Government investment should focus on groundwater recharge mechanisms such as percolation ponds at village-level. At farm level, provision of subsidized, long-term credit for well digging and pumpsets is important since farmers in rainfed area lack investment capacity. Speedy provision of electricity connection for agricultural pumpsets in these areas is also important. Subsidies for well digging may be linked with compulsory adoption of drip and sprinkler irrigation technologies and farm ponds.

(II) Dryland areas with poor groundwater potential

Subsidized credit and crop insurance are essential to enable the farmers make investments in modern crop production technologies and to mitigate the risks in rainfed agriculture. These programmes may be linked with compulsory adoption of on-farm soil and water conservation investments. A part of the cultivation loans must be set aside for investing on these structures. Government investment should focus on soil and water conservation structures at village level. Fodder crop and livestock production is a very important component in resource poor regions so as to supplement crop income, provide cheaper soil nutrients and to reduce crop risks. Fodder availability in dry seasons/drought years may be an important constraint in promoting livestock production in these areas. Hence, creation of community fodder reserves/fodder buffer stocking is an important pre-requisite to make livestock husbandry sustainable in this type of production environments. Drought-tolerant fruit and tree crops should be promoted to provide supplementary income to farm households.

(III) Dryland areas where people have diversified their livelihood away from agriculture

As indicated by the trends in population growth in rainfed areas, there has been a significant out-migration from rural areas of rainfed regions most probably due to lack of livelihood opportunities in these areas. This is also reflected in increasing trends in fallow lands in most of the rainfed districts. Thus, the policy for these areas should focus on less-labour and management-intensive crops.

- Drought-tolerant tree and fruit crops that require less labour inputs and day-to-day care and management.
- Rain water harvesting to provide supplementary irrigation to tree crops

- Provision of subsidy for labour-saving implements and machinery

Is subsidy for field operations workable?

Subsidies for field operations are unlikely to work because of the problems of cheating and bogus claims. Further, these kinds of subsidies are costly to administer, because of recurring nature of these operations. Therefore, a possible alternative is to provide capital subsidy for community ownership of soil and water conservation and labour-saving equipments and linking these with rainfall insurance. Cost-effective rainfed farming technologies could be demonstrated and disseminated through demonstration in farmers' fields by following the model of precision farming project being implemented in the state.

Recognition of risk as a major hurdle in promoting rainfed farming is a major step to derive appropriate policies to develop rainfed agriculture. There is a need for time bound programme to deliver credit and insurance services through group credit and group insurance to overcome default and moral hazard problems.

Future rainfed agricultural development policy should focus on following aspects:

- Technology: Major technological interventions in rainfed farming shall include farm-level soil and water conservation, new enterprise mix to suit specific rainfed production environments, new crops and crop varieties, and package of practices.
- Livestock is an integral component of rainfed enterprise mix because livestock are insurance against crop risks and also provide cheaper draft power and soil fertility. Therefore, any rainfed farming programme should recognize the fact that suitability of crop residues as livestock fodder as a key factor determining farmers' crop choice in rainfed areas.
- Institutions – Credit and insurance are the important institutional requirements to promote dryland agriculture. Self-help and community initiatives could

be promoted for provision of credit and insurance and also low-cost, labour-saving implements and machinery on custom hiring basis.

- Infrastructure – Roads and small-scale soil and water conservation structures and agro-processing (for dryland crops) are the important infrastructure requirement for rainfed areas. Value addition for dryland crops at village level will not only increase the profitability of rainfed crop production but also create employment.

Contributors

L.M. Ahire, Assistant Professor, Kerala Agriculture University, Kerala.

Sandhya Shenoy, Principal Scientist, National Academy for Agriculture Research, Hyderabad.

V.K. Jayaraghavendra Rao, Senior Scientist, NAARM, Rajendranagar, Hyderabad-500030, Andhra Pradesh.

S.B. Shinde, Associate Professor of Agricultural Extension, Department of Extension Education, Mahatma Phule Krishi Vidyapeeth, Rahuri-413722, Distt. Ahmednagar, Maharashtra State.

A.G. Sawant, Former Member, ASRB, ICAR, New Delhi.

G.J. Patel, Narmda Irrigation Research Project, Anand Agricultural University, Khandha, Gujarat.

D.B. Patel, Narmda Irrigation Research Project, Anand Agricultural University, Khandha, Gujarat.

T.P. Kotecha, Narmda Irrigation Research Project, Anand Agricultural University, Khandha, Gujarat.

K. Thangaraja, SRF (Agrl. Extension), DAE & RS, TNAU, Coimbatore.

C. Karthikeyan, Associate Professor (Agril. Extension), TNAU, Coimbatore.

J. Venkatapirabu, Associate Professor (Agril. Extension), TNAU, Coimbatore.

K. Chandrakandan, Emeritus Scientist, ICAR, Department of Agrl. Extension, TNAU, Coimbatore.

C. Cinthia Fernandaz, SRF, Emeritus Scientist Schemes, Department of Agrl. Extension, TNAU, Coimbatore.

R. Ganesan, Professor, Department of Agrl. Extension, TNAU, Coimbatore.

P.K. Mishra, Head, CS&WCRTI Research Centre, Bellary, Karnataka. Email: pkmbellary@rediffmail.com

M. Osman, Senior Scientist (Agronomy), CRIDA, Hyderabad, Andhra Pradesh.

G.R. Maruthi Sanker, Principal Scientist, CRIDA, Hyderabad, Andhra Pradesh.

Asim Biswas, Department of Soil Science and Agricultural Chemistry, University of Agricultural Sciences, GKVK, Bangalore, Karnataka.

S.L. Budihal, National Bureau of Soil Survey and Land Use Planning (NBSS&LUP), Bangalore, Karnataka.

P. Balaji, Department of Agricultural and Rural Management, Tamil Nadu Agricultural University, Coimbatore, Tamil Nadu.

C. Velavan, Department of Agricultural and Rural Management, Tamil Nadu Agricultural University, Coimbatore, Tamil Nadu.

G.D. Singh, Director Extension, Rajasthan Agricultural University, Bikaner.

M. Shantha Sheela, Assistant Professor, Agricultural Extension.

R. Rajasekaran, Assistant Professor, Agricultural Extension.

K. Palanisami, Director, Centre for Agricultural and Rural Development Studies, TNAU, Coimbatore, Tamil Nadu.

D. Karpagam, Directorate of Planning and Monitoring, TNAU, Coimbatore, Tamil Nadu.

E. Mahalakshmi Geetha, Junior Research Fellow, CARDS, TNAU.

N. Kalaichelvi, Junior Research Fellow, CARDS, TNAU.

A. Manickavalli, Junior Research Fellow, CARDS, TNAU.

V.K. Jayaraghavendra Rao, Senior Scientist, Agricultural Extension.

N. Sandhya Shenoy, Principal Scientist (Agril. Extn), NAARM, Rajendranagar, Hyderabad, Andhra Pradesh.

A. Raviraj, Research Scholar, Department of Agricultural Economics (DAE), CARDS, TNAU, Coimbatore.

S. Gurunathan, Associate Professor (SWCE) in Water Technology Centre, TNAU, Coimbatore.

S. Gurunathan, Department of Agricultural and Rural Management, Tamil Nadu Agricultural University, Coimbatore.

M. Malini, Department of Agricultural Rural Management, CARDS, TNAU, Coimbatore.

M. Ranichandra, Department of Agricultural Rural Management, CARDS, TNAU, Coimbatore.

Ashok Kumar, Department of Agricultural Economics, CSK HPKV Palampur (H.P.).

K.C. Sood, Department of Agricultural Economics, CSK HPKV Palampur (H.P.).

Atul Dogra, Department of Agricultural Economics, CSK HPKV Palampur (H.P.).

S.K. Sharma, Department of Agricultural Economics, CSK HPKV Palampur (H.P.).

N. Deepa, Senior Research Fellow, Department of Agricultural and Rural Management, TNAU, Coimbatore.

A. Rohini, Assistant Professor, Department of Agricultural and Rural Management, TNAU, Coimbatore.

S. Moghana Lavanya, Assistant Professor, Department of Agricultural and Rural Management, TNAU, Coimbatore.

K. Bassi, Directorate of Extension Education, CSK HPKV, Palampur (H.P.) 176062.

S.S. Masand, Directorate of Extension Education, CSK HPKV, Palampur (H.P.) 176062.

D.S.M. Gowda, Krishi Vigyan Kendra, Hanumanamatti-581135, UAS, Dharwad, Karnataka.

S.M. Hiremath, Krishi Vigyan Kendra, Hanumanamatti-581135, UAS, Dharwad, Karnataka.

P.V. Kusagur, Krishi Vigyan Kendra, Hanumanamatti-581135, UAS, Dharwad, Karnataka.

K.B. Yadahalli, Krishi Vigyan Kendra, Hanumanamatti-581135, UAS, Dharwad, Karnataka.

H.R. Nagaraju, Krishi Vigyan Kendra, Hanumanamatti-581135, UAS, Dharwad, Karnataka.

R.A. Budihal, Extension Co-ordinator, UAS, Dharwad, Karnataka.

G.P. Reddy, Sr. Scientist (Agril-Economics), National Academy of Agricultural Research Management, Rajendranagar, Hyderabad.